Oryx Frontiers of Science Series

RECENT ADVANCES AND ISSUES IN COMPUTERS

by Martin K. Gay

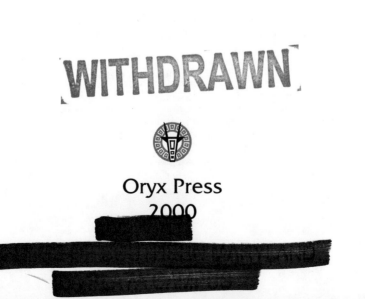

Oryx Press

2000

The rare Arabian Oryx is believed to have inspired the myth of the unicorn. This desert antelope became virtually extinct in the early 1960s. At that time, several groups of international conservationists arranged to have nine animals sent to the Phoenix Zoo to be the nucleus of a captive breeding herd. Today, the Oryx population is over 1,000, and over 500 have been returned to the Middle East.

© 2000 by Martin K. Gay
Published by The Oryx Press
4041 North Central at Indian School Road
Phoenix, Arizona 85012-3397
www.oryxpress.com

Published simultaneously in Canada
Printed and bound in the United States of America

∞ The paper used in this publication meets the minimum requirements of American National Standard for Information Science—Permanence of Paper for Printed Library Materials, ANSI Z39.48, 1984.

Library of Congress Cataloging-in-Publication Data

Gay, Martin, 1950–
   Recent advances and issues in computers / Martin K. Gay.
      p. cm.—(Oryx frontiers of science series)
   Includes bibliographical references and index.
   ISBN 1-57356-227-0 (alk.)
   1. Computers. 2. Computer science—Research. I. Title. II. Series.
QA76.5.G355 2000
004—dc21   00-035679

For Dick Barnhart,
one of the people ensuring that the digital revolution is about
enhancing our opportunities to enjoy each other

CONTENTS

PREFACE

*R*ecent Advances and Issues in Computers is one volume in the Frontiers of Science reference book series published by The Oryx Press. Concentrating on the core disciplines, the works in this series are designed to provide a brief but thorough overview of the latest developments, trends, and innovations in science research and technology with additional emphasis on how to best prepare to enter these challenging fields of study.

The coverage of this subject area will be somewhat different from the others in the series in that "computer science" is not recognized as a distinct science. In fact, the concept of computer science is generally agreed to have arisen only at the beginning of the 1960s, with the first digital computer being developed just 20 years prior. The advances in technology that made possible this breakthrough device were a direct result of the research efforts of electrical engineers and physicists (e.g., designing circuits that could produce arbitrary outputs, inventing the transistor, and perfecting storage media) and theoretical mathematicians (e.g., developing Boolean algebra and binary data theory). Computer science, as it is taught and practiced in universities and research centers today, has emerged from the synthesis of these and other disciplines. In the simplest terms, it is the study of the technology behind the information machines that are changing the world.

Within this discipline, one will find computer engineers, who design computing machines, and computational scientists, who develop the scientific, academic, and commercial applications that ever-more-powerful computers are capable of running. Mathematicians continue to contribute theoretical insights that often drive the empirical research of both computer engineers and computational scientists. In addition, the new field of information science, which has evolved from library science, psychology, engineering, and linguistics, studies how computers can be best utilized for the collection, storage, organization, retrieval, and use of information.

Regardless of how the technicians, engineers, scientists, or researchers are identified, their work in computer technology can be generally described in terms of one of the three subdisciplines of computer science. These subdisciplines include

- Architecture. This area includes all levels of hardware design and the interface between the equipment and the software that makes it a complete system.
- Software. The programs that direct computer operations are developed in this area. Software engineering, artificial intelligence, programming languages, operating systems, information systems, databases, and computer graphics are all subsets of the software concentration.
- Theory. Work on computational methods, numerical analysis, data structures, and algorithms characterizes computer theory.

The advances in this field have been driven by a practical need to create and/or improve machines that will (1) facilitate and accelerate knowledge gathering in all of the sciences and (2) aid business in the process of data management. This book will provide an overview of how those objectives have been met between 1996 and the present. In "computer years," this time period is almost a blur of faster and better, but the freeze-frames of the digital landscape presented here should prove helpful for the serious student contemplating a career in computers, or for the knowledgeable lay reader trying to make sense of the information-technology-driven changes taking place in all levels of society.

Information for this book was collected almost exclusively from Internet-based resources found around the world. The author culled the most relevant and reliable information from authoritative sources and organized it into a meaningful and coherent overview of the recent advances and issues in computers. Where appropriate, URLs have been included so that the reader may continue an inquiry into a particularly interesting or complicated subject area.

As most of us have come to discover, hyperlinks are not forever, and readers should be prepared to do some intelligent investigation based on the home domains referenced. Another excellent technique for tracking down related material on the Web is through two particularly good search engines: www.hotbot.com and google.com.

STRUCTURE AND CONTENT

This book is organized under the same general chapter headings found in the other books in the Oryx Frontiers of Science Series with the exception of a few chapter headings that differ because computer science doesn't always lend itself to the categories of the traditional sciences. The content of the 11 chapters in this book is described below.

Chapter 1 concentrates on recent advances, developments, and discoveries that have the potential for improving or changing the way computers operate. Topics in this section include some of the groundbreaking work being done on new materials and procedures that have the potential to replace the current CMOS (complementary metal oxide semiconductor) architecture of the silicon-based microprocessor, changes in the methods now used for mass storage systems, and a look at emerging new forms of computing environments.

Much of the motivation behind the research at the country's leading supercomputing centers is coming from the Herculean effort to build the new Internet. Chapter 2 describes the national computational grid for high-bandwidth, high-speed collaboration among scientists, which is expected to evolve into a prototype for the type of connectivity the entire country will enjoy in the coming decade. This chapter will look at the work now being undertaken by the two consortia charged with implementing this infrastructure, namely, the National Computational Science Alliance (the Alliance), led by the National Center for Supercomputer Applications at the University of Illinois; and the National Partnership for Advanced Computational Infrastructure, led by the San Diego Supercomputer Center at the University of California, San Diego.

Chapter 3 examines recent product developments in the computer and information technology industries which are influencing the way businesses and individuals interface with their world. Since no one book could adequately cover all of the developments, or guess what the most important releases in hardware and software in this area are, the focus in this section will be on trends and movements in the industry. Networking, portability, and human-computer interfaces are included.

Chapter 4 looks at some of the current social, political, ethical, and economic concerns that have grown around the developments in computer science and the technological capabilities outlined in the previous three chapters. Microsoft's influence, as well as privacy, encryption, universal access, adaptive technologies, and gender issues, are a few of the concerns that are discussed.

Chapter 5 provides brief biographical outlines of important individuals who were instrumental in the developments and/or issues documented in chapters 1-3. In some cases, these are not the most famous names in computing, but they are included here because their work in computer architecture, software, or theory has influenced the direction of the technology in recent years.

Chapter 6 contains documents and speeches that provide background and understanding of some of the important issues and developments in the computer field.

Chapter 7 provides a snapshot of the career paths available in the computer and information services industries, with a focus on education, skill, and experience requirements. This chapter takes a brief look at the careers one would traditionally associate with information technology (IT) employment and it identifies some new categories that have recently emerged because of the explosion of Internet-driven opportunities.

Chapter 8 displays graphs, tables, and maps that focus on careers and trends associated with computing and the development of new applications and information systems.

Chapter 9 lists organizations and associations, both national and international, dedicated to computer science, computational science, computer engineering, or issues related to the way IT is changing society.

Chapter 10 is an annotated bibliography of important reference books, some technical texts, journals, and key World Wide Web sites.

Chapter 11 provides a glossary of terms useful to the basic understanding of computers and their related systems. The excerpts presented here are from Denis Howe [editor of *The Free On-line Dictionary of Computing* (FOLDOC), at wombat.doc.ic.ac.uk], and they have been edited slightly for space and to update some of the terms.

CHAPTER ONE
Basic Computer Research Today

The computer science research and development departments at the top universities and corporations have been engaged in thousands of projects in the past three years. While results from these many institutions can be grouped into dozens of areas of inquiry, it is useful for the purposes of this book to key on a few of the most critical ones. The overall trend in computing is toward the eventual development of a digital infrastructure that will touch every corner of human existence. And while we are still in the infant stage of this process, the speed at which changes are occurring leads many experts to predict that visions like the "digital nervous system," Microsoft's term for a broadly distributed and easily accessible computer network, will be in place sometime in the first decade of the twenty-first century.

To implement this vision, several conditions have to be met. These include

- More powerful microprocessors
- Lower cost and smaller components
- Improved human-machine interfaces
- Ubiquitous, broad-band connectivity
- Robust user security to assure privacy
- Transparent operation
- Simplified systems

The dramatic advances and possibilities discussed in this chapter have been chosen because they show significant promise toward meeting the challenge of these predicate factors. The first area covered is microprocessor development. Without question, changes to this nerve center of the computer have had the most significant impact on the capabilities of the smart machines. "Smaller and faster," the computer industry mantra, is a direct result of the microprocessor shrinking in size and expanding in power.

The focus then shifts to mass storage options and the surprising research being carried out on the use of new media and protocols (compression modalities and access techniques) for the stable maintenance of both short- and long-term data. This section is followed by discussions on the development of new human-machine interfaces that have become possible due to improvements in microprocessors and programming. Finally, the change in the capabilities of the most powerful of the information appliances, the supercomputers, concludes this brief overview of the recent inroads in computer science research.

THE AGE OF MICROPROCESSORS AND INTEGRATED CIRCUITS

Because the integrated circuit is the brain inside the computer and virtually every other device driving the Information Age, improving its efficiency and power has been a top priority for hundreds of research centers around the globe.

The basic configuration of these microprocessors, or chips as they are called, has not changed since they were first conceived by Gilbert Hyatt in 1969. Hyatt's work built on the groundbreaking research of Texas Instrument researcher Jack St. Clair Kilby, who is credited with development of the first transistor, a solid state device that replaced vacuum tubes as a method to control the on/off state of electricity moving through a medium. Hyatt never capitalized on his invention, however, and in 1971, the newly formed Intel Corporation began manufacturing and marketing the first commercial microprocessors.

Consisting of a single structure with no connecting wires, the device is an assembly of semiconductor elements (made of crystalline material which allows current to flow under certain circumstances), including transistors and diodes, and passive arrays of resistors and capacitors. These electronic devices, connected via a superfine line of aluminum, have been constructed on a common facilitating material, a substrate of silicon. Transistors supply the active microscopic switches on the substrate, and they have traditionally been formed by adding elements like arsenic or boron to the base material. These impurities alter the manner in which electrons move through the silicon (Texas Instruments 1999).

A thorough discussion of the history of integrated circuits and microprocessors and of how microprocessors are manufactured is provided by the World Wide Web (Web) sites maintained by Texas Instruments (www.ti.com) and Intel (www.intel.com). In short, after an ultra-thin slice of silicon, approximately 1 square inch in size, is polished and readied in a facility that is cleaner than any hospital operating room on earth, chips undergo eight key processing steps:

1. Deposition, or growing an insulating layer on the slice of silicon, installs a layer on the wafer's silicon substrate which can be patterned using photolithography to form circuit elements.

2. During diffusion, impurities are baked into the wafer in a diffusion furnace. Electrical characteristics are thereby altered to create separate regions with excess negative or positive charges.

3. Metallization is a type of deposition process. Here, many interconnections are formed on each of hundreds of integrated circuits on the wafer. Metallization is also used to make the bond pads that interconnect a chip to other components on a printed wiring board.

4. During ion implantation, dopants or other impurities are introduced into a wafer's surface to create gate oxides, traditionally from silicon crystals that conduct and control electron flow.

5. Photolithography patterning refers to creating the actual circuitry. A chemical coating called photoresist is applied to the wafer surface before exposing the coated area to ultraviolet light. The photoresist hardens in desired patterns when it is developed.

6. During the etching process, wafers are moved to a plasma reactor where electrically excited gases etch the surface into the pattern that was defined by the photolithography process. After etching, wafers are cleaned thoroughly.

7. Near final wafer fabrication, each wafer is subjected to testing to detect defective components.

8. Silicon nitride, a protective coating, is applied. Wafers are then ready for the final processing step, multiprobe testing. Each integrated circuit in the wafer (now a microprocessing chip), is electrically tested to determine whether or not it is ready for final assembly, bonding, and packaging.

Moore's Law and the Emerging Paradigm

More transistors translate into greater performance for the chip. When Robert Noyce and Gordon Moore joined Andy Grove in 1971 to form the integrated circuit manufacturing company, Intel, their first product was a 1-kilobyte, dynamic-random-access memory (DRAM), integrated circuit; it was the first chip capable of storing a significant quantity of data. But in 1974, they created an even more important product, the first microprocessor, which International Business Machines (IBM) would eventually license as the central processing unit (CPU) for its new personal computer (PC). That chip contained 2,300 transistors. By 1997, the Pentium II processor manufactured by Intel boasted 5.7 million transistors. The Pentium III-Xeon was introduced in mid-1999. It came with 9.5 million transistors.

The increase in processing performance has generally followed the developmental curve outlined in what has come to be known as Moore's Law. In 1965, Gordon Moore had noted that the number of transistors in integrated circuits had doubled approximately every 18 to 24 months. And that formula has held remarkably true up through the newly released Pentium III product.

The Semiconductor Industry Association is responsible for *The Technology Roadmap for Semiconductors (TRS),* an overview and a predictor of how to maintain the steady pace of growth in computing power that has been the hallmark of Moore's Law. It attempts to set benchmarks through the year 2012 that will affect standards and fabrication of DRAM bit count (storage capacity), logic performance (input speed), and mixed signal integration through the use of ever-decreasing geometry on silicon-complementary metal-oxide semiconductor (CMOS) integrated circuits. That is, this effort of the chip industry has sought to document and distribute the best practices for making transistors smaller so that microprocessors become more efficient and powerful.

The greatest challenge to continuing the smaller, faster, cheaper scenario in CMOS chip development noted in the *TRS* is the 0.10 micron barrier. (One micron is one-millionth of a meter.) *TRS* predicts this wall, at which electron pathways cannot be made narrower, will be reached around 2006. This barrier is due to the physical limitation in tools (visible light emitting techniques) that can create circuits under the 0.10 micron level. The standard process used to imprint paths in silicon wafers today is optical projection lithography, which achieves a standard 0.25 micron. After that, to reach the ranges of 0.10 micron to 0.05 micron that the *TRS* anticipates will be necessary for 2012, a number of new techniques are being considered. As a reference point, today's common silicon-based chip has transistors that are 0.25 microns. This size was brought down to 0.18 microns in 1999, two years ahead of what the *TRS* has predicted. But, before any company makes the substantial investment in fabrication systems and plants that will be required to reach and break the 0.10 micron barrier, it will have to be convinced that the process is likely to result in a reliable end product. For that reason, research is moving apace in many directions. The size of the transistor's path, of course, is an overriding consideration, and innovative new imprinting techniques are being adopted. Other factors are being rethought as well:

- Replacing the aluminum connections between semiconductors with copper or a new experimental material called a xerogel might improve speed. These and other materials have shown significantly lower delays in making connections.
- Building the chip itself out of new material or improving the silicon with an application like IBM's silicon-on-insulator (SOI) might be more effective. This process places a thin layer of silicon (which then holds the CMOS circuitry) on an insulating material like silicon oxide or glass. RCA Solid State has also had some experimental success using a sapphire (alumina) substrate (SOS) in the same manner.
- Bulk applications that would replace silicon completely are being considered. Though more expensive and less stable than the traditional chip material, silicon germanium (SiGe) and gallium arsenide (GaAs) are two such compounds under investigation.
- Materials and techniques that can provide electron connectivity without the usual accompanying heat build up could be developed.
- Improving the fabrication processes already in place by tightening tolerances and improving sterile conditions in fabricating plants could result in significant progress. Intel, for instance, spent $5 billion in 1999 just to improve and adapt manufacturing to reach the 0.18 micron level.
- The use of light spectrum tools to produce chips could be abandoned, because as wavelengths become too short, they turn into X-rays and damage molecules in the process. Instead, some experimental work at Hewlett-Packard (HP) has shown that by building components chemically, circuits can be shrunk to very small levels. Their Teramac computer has run 100 times faster than the other most powerful HP workstation available commercially. Because the chemical connections can also build-in alternative paths that allow connections to find another method to get

a signal from here to there, the new chip does not have to be perfect. That means they will be a lot less expensive to fabricate.

- The possibility of setting up Biological computation systems that utilize biologically grown neurons, which are modified for a particular computing environment, is being investigated. Appropriate uses would be in brain-like systems where pattern recognition, natural language, and control of motion is important.

Gate Oxide Advances

One important strategy for speeding up transistor operation is improvement of the gate oxide layer. On present day microprocessors, this fine film of material is typically composed of silicon dioxide. It serves the important role of insulator, or gatekeeper, between the gate electrode and the channel through which electric current flows. If the gate oxide is made thinner than the 20-atom width now common in 0.25-micron transistors, a higher voltage can pass through the gate and result in faster switching speeds. Unfortunately, as the gate oxide thins, its behavior becomes less predictable, with deterioration and failure becoming concerns.

Much of the development of this chip technology has been centered at one facility in particular, Lucent Technology's research arm, the famous Bell Labs. In some of the latest efforts there, researchers have provided some insight into the attributes of the gate oxides. They have developed a software program that can accurately predict how much energy can be passed through a gate, and where and when degradation will appear.

In a related development at the University of Delaware, Electrical Engineering Professor James Kolodzey and his colleagues have announced a new technique for producing alumina gate oxides that provide an electrical storage capacity three times greater than silicon dioxide. If alumina gate oxides can be successfully combined onto the silicon substrate, they can actually be made three times thicker, and hence, more reliable than the commonly used element. It is expected that they could also increase speeds and performance for certain functions by as much as 1,000 times.

This work coincides with another Bell Labs breakthrough that demonstrates a method by which variable thicknesses of gate oxide can be applied to a chip's surface. Once again using silicon dioxide, the material is "grown" on the surface of the substrate by adding oxygen to the silicon surface. That growth can be retarded and thereby controlled to allow variable thicknesses, through the addition of nitrogen. To test this process the Bell Labs researchers created a high-performance test circuit with 0.18-micron features and multiple gate-oxide thicknesses. Even though the same level of power was applied, the electron speed in this circuit proved to be twice as fast as regular uniform thickness gate oxides.

System on a Chip

One of the consequences of using variable-thickness gate oxides would be improved performance for the new system-on-a-chip products that are taking integrated circuits to a new level of flexibility and performance. In the past couple of years, manufacturers of digital products like cellular phones have found it to be a great advantage to combine several functions on one chip. Memory, digital logic and analog circuits, and input/output functions can now be incorporated as various components of a single microprocessor. The gate oxide is of the same thickness for these diverse integrated components and this causes some to work less efficiently than others. If variable thicknesses of gate oxide can be applied so as to maximize the performance of each function, then speed, reliability, and overall performance of the product can be enhanced. Single-chip cellular phones the size of a wrist watch are on the horizon. But a single-chip PC is already here.

Figure 1.1. Merlin Scanner-on-a-Chip (on finger) performs all the functions that were previously handled by large, bulky, printed circuit boards that used as many as 40 integrated circuits.

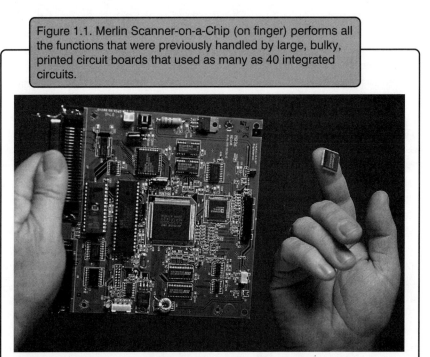

Courtesy of National Semiconductor Corporation. Source: http://www.national.com/company/pressroom/gallery/dcs.html.

Introduced in 1999, National Semiconductor created the industry's first single-chip PC. This single microprocessor, less than 0.5 inch in width, contains at least 12 separate component functions. The one chip will replace all of those separate chips now designed into most PCs on the market. Current plans call for the creation of this low-priced chip on the current 0.25-micron architecture, with a move to the 0.18-micron architecture in the next few years.

National Semiconductor CEO Brian Halla said, "First the PC goes on a chip. Next, the PC becomes a plug-in behind the dashboard of your car, behind a flat-panel display in your kitchen, or inside a set-top box. The PC disappears just the way electric motors are invisible in our lives. We use them all day long, but we only think about the appliance, not the motor" (DeTar 1998). Prior to this prediction, Halla had already announced that his company had perfected a complete scanner-on-a-chip design.

Beyond the Silicon Wafer

Bell Labs made news again when it recently advanced a 30-year effort to create a stable, gallium-arsenide-based, metal-oxide-semiconductor-field-effect transistor (MOSFET). This effort follows up its successful 1997 test, in which molecular beam epitaxy was used to deposit atomic-sized layers of the ultra-thin gate-oxide material on the gallium-arsenide channel surface.

The channel is the active region of the chip structure. It is in this region where the electrons flow from the source to the drain outflow. It is imperative that the connection between the gate oxide and the gallium-arsenide channel be as free from defects as possible, because the gate acts as an insulator. Defects and nonsmooth contact between these layers can lead to instability, i.e., short circuit and failure. That is where the challenge resides, and this is specifically where the latest test has shown results. In the 1998 research, Bell Lab scientists were able to improve the stability of the gate oxide (made of gallium oxide and gadolinium oxide) to extend the stable operation of the resultant chip for several hours.

Ideally, this device would replace the silicon-based MOSFETs now used in signal processors and memory modules because of the increased speed they afford; however, the first application is likely to be in cellular phones and communication systems. These devices currently use chips based on the gallium-arsenide channel, but without a gate oxide. They require higher frequencies than silicon chips can deliver. But, these higher frequencies can be delivered by gallium-arsenide chips, because electrons move six times faster down channels made of this material.

Another promising procedure being touted by a team of electrical engineers from Mississippi State University is using silicon carbide, an adaptation of the common silicon chip material. Adding carbon to silicon, they contend, results in a better product. Because it is considerably stronger than silicon alone, it is especially useful in high-temperature, high-voltage, and high-frequency applications. The main problem with silicon-carbide semiconductors is the cost. Even though they have proven to be more reliable, they can cost a thousand times more than their common cousin. Current research is looking at how these cost differences can be narrowed.

Nanotubes and Nanotechniques

One important area of research that could eventually lead to the development of very small processors and tiny computing machines is nanotechnology. Ralph C. Merkle, winner of the 1998 Feynman Prize in Nanotechnology, is a Xerox Palo Alto Research Center (XEROXPARC) researcher and one of the foremost authorities in the field. He is also executive editor of the journal *Nanotechnology* and a director of the Foresight Institute. Dr. Merkle maintains the nanotechnology Web site (nano.merkle.com/nano), where he defines molecular nanotechnology in this clear language:

> Manufactured products are made from atoms. The properties of those products depend on how those atoms are arranged. If we rearrange the atoms in coal, we can make diamond. If we rearrange the atoms in sand (and add a few other trace elements), we can make computer chips. If we rearrange the atoms in dirt, water, and air we can make potatoes.
>
> Today's manufacturing methods are very crude at the molecular level. Casting, grinding, milling, and even lithography move atoms in great thundering statistical herds. It's like trying to make things out of LEGO blocks with boxing gloves on your hands. Yes, you can push the LEGO blocks into great heaps and pile them up, but you can't really snap them together the way you'd like.
>
> In the future, nanotechnology will let us take off the boxing gloves. We'll be able to snap together the fundamental building blocks of nature easily, inexpensively, and in almost any arrangement that we desire. This will be essential if we are to continue the revolution in computer hardware beyond about the next decade, and will also let us fabricate an entire new generation of products that are cleaner, stronger, lighter, and more precise (Merkle 1999).

Practical Uses of Nanotechnology

Professor Z. L. Wang of the School of Materials Science and Engineering at the Georgia Institute of Technology and his colleague, Professor Walter de Heer, have done some preliminary research on the properties of newly discovered carbon nanotubes. These single-walled carbon molecules, discovered in the early 1990s, are believed to be created from an electrical discharge between electrons. They display a property known as ballistic conductance, which means that electrons pass through them without a heating effect at room temperature. Dr. Wang noted

> In classical physics, the resistance of a metal bar is proportional to its length…. If you make it twice as long, you will have twice as much resistance. But for these nanotubes, it makes no difference whether they are long or short because the resistance is independent of the length or the diameter. The electrons are passing through these nanotubes as if they were light waves passing through an optical waveguide…. It's more like optics than electronics (Science Daily 1998).

The significance of ballistic conductance for computer construction is obvious when one looks at previous attempts to create microcircuits with superconductive properties. Because faster, smaller connections generate too much heat in traditional CMOS-designed chips, performance gains due to architectural changes cannot be sustained unless a supercooling mechanism is attached.

One successful commercial application was the KryoTech solution announced by Digital Equipment in 1998 for its 767 Personal Supercomputer. Researchers there have been able to increase the performance of the processor by 30% by including the KryoTech company's supercooling system. The attached refrigeration unit kept the CPU at minus 40 degrees Celsius. Eliminating the need to cool the connection in the first place will be a quantum step in the development of truly microscopic devices with more power than is considered practical today.

With the advent of a practical configuration of nanotubes or a related carbon-based switching mechanism, computers could be functional at the molecular level. They would be millions more times powerful than today's supercomputers. Experts are looking toward the year 2010 or beyond for a practical application of these principles.

In another research venue, scientists at the Weizmann Institute have provided one of the practical answers regarding how this new nanotechnology might be applied. By using the chemical theory of liquids, these researchers found a way to predict the minimum possible size of transistors commonly used in microelectronics. Next, they actually built a tiny prototype using the experimental semiconductor copper indium diselenide. The inner core of this transistor was 20 nm (with a total overall width of 50 nm). That made it five times smaller than current products.

This feat was accomplished using atomic-force microscopy, which uses a microscopic stylus to enter the material and then manipulate atoms in the semiconductor. When a voltage was applied to the stylus, it caused atoms that determine an element's conductivity (known as dopants) to be directed in a particular pathway. In the experiment, only 100 to 200 dopants were moved. This small number, however, was enough to produce a tiny transistor.

The effects of nanotechnology and nanoscience, while perhaps not obvious today, will dominate the breakthroughs of the future. When asked from what horizon the next world-changing innovation will appear, National Science Foundation Director, Neal Lane answered, "I would point to nanoscale science and engineering, often called simply 'nanotechnology.'"

BEYOND BINARY COMPUTING AND THE TURING MACHINE

An interesting area of research, and one that was thought of as more theoretical than practical for many years, is the possibility of constructing hardware and software programs for "quantum computing" and "chaos computing." Up until this point, the model of computation has been mathematician Alan Turing's concept, as proposed in his breakthrough theoretical paper "On Computable Numbers with an Application to the Entscheidungsproblem." It was presented to the Proceedings of the London Math Society in 1937. That model is familiar to most readers today as it stipulated a device that would use various programmable components, represented by binary digits that would exist in a state of on or off (1 or 0), to perform various unrelated tasks. Current chip technology with its system of millions of transistors on integrated circuits is what has developed in seeking to duplicate the Turing Machine.

Quantum Computing

The quantum computer is not limited to expressions in bits of data. Scientists have posited that it could function on the principles of quantum mechanics. Here, the smallest unit of information is a quantum bit (qubit), based on atomic particles like photons or ions trapped in an electromagnetic field. And because of the wave nature of the quantum world, qubits would actually contain a much greater amount of data than the bits now available. In 1994, Peter Shor of AT&T Labs defined a quantum algorithm for factoring integers. He proved that it would run exponentially faster than any classically designed algorithm. Two years later, Lov Grover of Bell Labs discovered a quantum algorithm for searching unsorted data in a database. His method increased the speed of accessing a probable answer by 5,000 times over current techniques. This improvement is because qubits are not just off or on, they can be in multiple states at the same time. Suddenly, researchers throughout the world are looking into quantum algorithms, and many are now ready to be tested along with the hardware that runs them.

Richard Hughes of the Department of Energy's Los Alamos Neutron Science and Technology Group showed that the use of quantam algorithms is feasible, but mechanical problems still need to be worked through before a quantum computer can be implemented. In 1998, his team constructed a simple quantum computing device that used single ionized atoms to function like a computer memory module. The device at Los Alamos consists of a string of up to eight trapped calcium ions and optical switches that direct a pulsed laser beam onto individual ions. The lasers are used to cool the ions to a state of rest in what is known as a trap. Logic gate operations, the basic control of any computer operation, are then accomplished by directing a pulse from a titanium-sapphire laser to the ions.

David Cory of MIT joined other researchers at Los Alamos, i.e., Raymond Laflamme, Wojciech Zurek, and Emanuel Knill, later that year to use nuclear magnetic resonance (NMR) to test some theories about the quantum computer's ability to check for errors. A major concern regarding this new computer evolves from the quirky nature of quantum mechanics. Many feel that computers based on quantum physics would generate far too many errors. Cory's group has been able to predict the errors accurately using the NMR technology. This means that in the future, researchers will be able to program an accurate number of programming checks to ensure maximum accuracy.

While it is very unlikely that quantum computers will replace current desktop designs, they could prove to be very efficient as manipulators of large data sets and in the building of accurate models of physical phenomena. It has been predicted that such computers will begin making small calculations in 2001.

Chaos Computing

A separate, interesting phenomenon in research on new computing paradigms was introduced at about the same time as the work being done on the quantum devices discussed above. Georgia Institute of Technology physics professor Dr. William L. Ditto, in collaboration with Sudeshna Sinha of the Institute of Mathematical Sciences in Madras, India, are responsible for a cutting-edge application that has evolved from chaos theory. They describe a dynamics-based computation that would rely on optical computing techniques that use extremely fast chaotic lasers and a silicon-neural-tissue hybrid circuitry. So far, the team reports successful operations on problems including addition, multiplication, Boolean logic, and finding the least common multiplier in a sequence of integers.

Chaotic elements are interlaced in a grid, which allows for a triggering of responses in neighboring mechanisms once an assigned threshold is reached in the first element. This naturally parallel system works like an avalanche according to Ditto. "We have the elements interconnected so that they respond to their neighbors like the avalanching that occurs when you pile grains of sand onto a sand pile. You allow the elements to avalanche and the system to evolve chaotically, then do the avalanching again until the system settles down to the right answer. It takes a couple of iterations for it to settle down" (Georgia Tech Research News 1998).

In this highly theoretical paradigm, the answers to a problem evolve in a way that mimics chaotic systems in nature. Ditto has done some calculations on an optical system that would use ammonia lasers to function as a dynamics-based computing environment. Predicting that chaotic computers will be optimal for operations other than those that are now easily accomplished using digital machines, scientists are looking into pattern recognition problems, which current machines are not efficient at solving, to test their theories.

OPTICAL COMMUNICATIONS

As suggested by the work done by Ditto and Sinha, one of the most promising areas of computer research is light-based transmission of signals. The concept being investigated in most cases involves replacing electrons with photons, a unit of electromagnetic radiation, as the mechanism that passes information within and among computers. As was noted in the discussion of microchips and integrated circuits, speed and power are achieved by packing ever greater numbers of pathways (transistors) onto the substrate of silicon. As the number of circuits through which the electrons can pass increases, the power goes up, but so does the amount of heat generated. Photons, which have been successfully used to transmit data in communications systems based on fiber optic technology (involving the transmission of light through glass fibers), are of vital interest because they would generate far less heat. More light pathways could be included without degrading the medium and eventually making the tool unworkable. The type of light emitted by lasers provides the most practical path for making the transition from electron to photon computing. Many challenges will have to be overcome to bring this concept to the desktop, but in the last couple of years, breakthroughs have been made in many key areas of research. The following are just a few of the most important.

Microlasers

The development of lasers, devices that use the natural oscillations of atoms or molecules to generate fine beams of light, initially took place in the research suites of the Bell Labs in the 1950s. A few years later, Yale University led the research that would produce the first gas laser. In 1994, Bell produced a laser based on quantum and chaos theory that is known as a quantum-cascade laser. By 1998, Bell and Yale had once more collaborated to produce a variation of the quantum-cascade laser. This new microlaser is a mere 0.05 millimeters in diameter and is one of the class of semiconductor lasers.

What researchers at the two institutions have been able to show is that these new laser pulses can be deployed in a bow tie pattern that emits light in four narrow beams. Because these beams can be controlled and directed, they hold a great deal of promise for use in very high speed optical-switching applications. New computers and new networks are two potential developments related to this technology that may be only years away.

Photonic Lattice

Another promising research path has recently netted positive results. For over a decade scientists have tried to develop a mechanism that could easily bend light beams without leaking them, so that their properties could potentially be used in systems like the optical computer. In 1998, the Department of Energy's Sandia National Laboratories announced that it had been successful in an attempt to interlock minute slivers of silicon into a lattice framework structure that may actually meet the requirement for bending, but not leaking light waves. Called a photonic crystal because of the way its internal structure repeats (just like a naturally occurring crystal), the lattice has been shown to manipulate light in the infrared range. In short order, the team at Sandia expects to have a structure that will test waveguides in the 1.5 micron neighborhood. This is in the range of most optically transmitted data.

The successful development of this photonic lattice structure at Sandia was built on the earlier groundbreaking work that was done in their micromachine lab. Scientists there were able to build microscopically small machines and some of the tiniest integrated circuits, using their surface-etched manufacturing technology. By following similar processes, the waveguide patterns etched into the silicon substrate in this experiment achieved a density factor of 10 to 100 times greater than what can be achieved using the gallium arsenide process. The latter process is far more expensive too.

Self-growing Plastics

Samson Jenekhe and his colleagues at the University of Rochester in New York have achieved a remarkable feat. They have found a way to get plastic molecules to grow into a crystalline form, coating a layer of glass one centimeter square. Jenekhe explained the growth in a statement that accompanied the 1999 report on this discovery.

> Much of nature is a product of hierarchical self-assembly, and humans are the example par excellence. Each of us starts as a single cell encoded with the information to guide our growth into a larger structure—a complete human being. Making materials that are on their own smart, intelligent and able to orchestrate their own growth marks the chemistry and polymer science of the future (OE Reports 1999).

Jenekhe reports that his photonic crystals form as hollow spheres stacked in a honeycomb pattern. These structures will pass light in a predictable, orderly manner and can be used as display components, and more importantly, as switches. By replacing the electron-controlled microprocessor with a light-driven switch, computing speeds will be increased in geometric proportions.

STORAGE DEVICES

As the amount of data in the world increases at an exponential rate on a daily basis, the need for stable, easily accessible, and inexpensive storage devices has become of paramount concern. Technological advances in manufacturing and new compression techniques have translated into great economies for the average user of computers. Where it used to cost thousands to obtain a gigabyte of storage space, by 1999, 10 gigabyte hard drives are becoming commonplace in PCs that cost $1,000–$1,500. However, as digital utilities become more powerful, their need for information continues to rise, and better, cheaper storage is still a key goal. New ways to provide for it are on the horizon.

Hard-disk memory is constructed by building successive strings of magnetic switches that can be on or off as a result of a change in the switch's magnetic polarity. The optimal switch is operational in a weak magnetic field, but with enough stability to exist over long periods without degradation. A problem that surfaces when scientists experiment with new forms of electronic switches has to do with the long-range nature of magnetic fields. To increase the mass of memory cells, scientists would like to pack them more tightly together. Previously, this was not possible because close proximity meant that a change in the polarity of one switch would alter the state of neighboring switches as well, making the array useless for storage. New developments may soon resolve this problem, however.

Molecular Magnets

Scientist at the Weizmann Institute in Israel reported in 1998 that they had developed a new molecular magnet that could work in a much more isolated state than previous types of switches. Their nickel-dichloride magnets appear to remain unchanged as adjacent magnets alter their polarity, and they show little effect when environmental factors are applied. This new class of magnetic materials is made of clusters of inorganic molecules. They are substantially smaller than the metal-organic compounds that have been used to create most molecular magnets heretofore. They are also different in molecular shape from the usual structure associated with these molecular magnets: some are fullerenes while others are nanotubes.

These magnets are constructed using a bottom-up approach. As the scientists add one atom at a time to the structure, the molecule then self-assembles into a spherical layer. By adding each one-molecule thick layer to another, researchers have the option of building switches that are exactly configured for individual processes. Their magnetic properties can be defined in a more absolute manner than was previously available.

MAGRAM

A radically different approach to increasing storage potential is being carried out at the University of Utah, where researchers have announced a major breakthrough in the development of a wholly new type of memory. This group of scientists has created a class of magnetic-field sensors called magnetic random access memory or MAGRAM. These MAGRAM cells utilize magnetic fields to store data.

This cutting-edge development is taking place in secret, but the developers are confident that the new memory will display rapid access potential like that of current RAM modules. The big difference will come when power is cut to the memory device. With current RAM chip technology, once the electrons stop flowing, the magnet switches move to their rest state. All data is lost. The new MAGRAM is completely nonvolatile, meaning that it will retain data like a conventional hard drive, have a large storage capacity, and also be easily and quickly accessed.

Organic Mass Storage

The Air Force Research Laboratory Information Directorate awarded Syracuse University researchers $2.1 million in 1998 to study a biological answer to the mass storage challenge. That project, "Protein-Based Optical Memory Development," is looking at the possibility of turning a San Francisco Bay bacteria into a storage medium. Based on an optical (photon) memory system, Syracuse researchers are designing a prototype that uses the organic protein *Bacteriorhodopsin*. This material has demonstrated light-absorbing properties. It is obtained by distilling bacteria that grows in salt marshes.

In the announcement of the award, Bernard J. Clarke, program manager of the Air Force Information and Intelligence Exploitation Division, explains the unusual research:

> When you fly into San Francisco and the Bay has a purple color, that's the bacterium in high concentration. We hope to use the protein from the bacteria as the active ingredient in a memory media that will allow us to store the equivalent of 100,000 books on a single source. The crux of the Syracuse University research will be how to encapsulate the protein so that it retains its qualities without drying up (Air Force 1998).

This research would actually be an investigation of an interim process that builds on the dramatic 1997 discoveries of Michael Heller, a physical biochemist at Nanogen in San Diego (www.nanogen.com/tech.htm). His work demonstrated the possibility of creating an optical storage media made up of synthetic DNA "potentially capable of storing the contents of all American research libraries on a single disk."

Heller used the synthetic DNA as a support structure to hold light-responsive molecules called chromophores. When light hits the chromophores, they glow. Binary data is held in these arrays of DNA by distributing the material in particular logical patterns that will illicit responses (binary 1) or no responses (binary 0) when light is applied.

Holographic Storage

The concept of using holographs to store data is over 30 years old. Researchers have always considered this medium a great potential source of storage because of its unique nature. Since they are naturally three dimensional, holograms can be encoded with data that are superimposed over other data. They also, in theory, would afford remarkably fast access times, because data could be retrieved in parallel architectural schemes that might lead to speeds of 1 million bits per access. Bell Labs has recently announced a breakthrough that should lead to this technology.

Researchers at the institute are expecting to fabricate a 5.25-inch disk that will hold 125 gigabytes of data. They are also predicting that, while it can be recorded on only once, it will be stable enough to last many years. A bonus feature includes access read times of 30-50 megabytes per second. Primary researchers, Dr. Kevin Curtis, Dr. Lisa Dhar, and others, have worked four years to reach this level of accomplishment. Now, by using materials and processes that are readily available, the team expects to produce a product using inexpensive photopolymers. The resultant storage media may cost as little as $10.00 per disk once commercially available.

ARTIFICIAL INTELLIGENCE

The interface between human and machine, the way we communicate with computers and the way they respond to us, has undergone significant improvements since the first graphical user interface (GUI) and the mouse were invented by researchers working at Xerox Corporation's Palo Alto Research Center (XEROX PARC) in the 1970s. The interaction is becoming more direct, slightly more intuitive, and substantially more useful as innovations like voice recognition products, ergonomically correct keyboards, laser pointing mice, 3-dimensional graphics, and higher resolution displays become commonplace. But the effort to make machines "think" and really behave like humans, is still years away.

Professor of computer science and codirector of the Knowledge Systems Laboratory at Stanford University, Edward Feigenbaum, is the person most responsible for developing the science that studies how to make machines think like humans—artificial intelligence (AI). Much of the software known as expert systems was developed or had its start in the laboratory he founded at Stanford in 1965. His goal then, and now, is to train computers to solve problems the way humans do: build on basic knowledge steps achieved through trial and error or via direct instruction.

Progress in this incredibly complicated area has been slow to date. While machines are starting to actually make some decisions based on limited knowledge areas, as one researcher noted, this thinking is at the level of a dumb insect. One can barely imagine the millions of common sense rules humans use to function in this world. For computers to have access to these rules, they all have to be translated into code they can use. The head of Cycorp, Douglas Lenat, is one of Feigenbaum's students at Stanford. He and his company have not even reached the halfway point in what he estimates to be a 40-year coding effort to get common sense rules into software. One should look to 2025 before a reasonably intelligent machine emerges.

Artificial Intelligence Research Directions

Other companies and universities are doing work on more manageable portions of the problem. One of these companies, Stanford Research Institute (SRI), was founded immediately following the end of World War II, and although it is only one of many companies committed to advancing AI, it serves to illustrate where much of AI research is directed. Today, SRI is responsible for advances in such areas as digital color compression technology that is modeled after the way a human retina encodes data, open agent architecture—an intelligent "agent" software that tracks human interactions, and speech recognition applications and software tools. All three of which have had an effect on current AI research.

SRI has divided its AI research into the following classifications.

Natural Language

Scientists involved in this division work in the following areas:

- Multimedia/Multimodal Interfaces involve work on ways to understand the optimal use of natural language in multimedia interfaces. Typical projects include studies of how pen-voice systems are used most effectively; implementation of interfaces that use handwritten, verbal, and gestural input; and multimedia communication between a user and intelligent agent software.
- Spoken language systems research strives to integrate linguistic processing with speech recognition to increase accuracy of speech recognition and to make it feasible to use in practical applications.
- Written language systems deal with interpreting information in written text. The research includes theoretical work on the use of inference and common sense knowledge in text understanding as well as applications like FASTUS, a very fast information extraction tool.

Perception

Research under this category includes experimentation and theoretical work in machine vision, expert systems, evidential reasoning, and virtual reality. More specifically, the programs can be broken down as follows:

- 3-Dimensional (3-D) modeling and interpretation deals with computational models of image interpretation that allow for analysis without knowledge of the actual environment.
- Analysis of range images builds on the groundbreaking programs in laser range finders at SRI.
- Image matching and autonomous navigation involve cartographic interpretation: matching images to map coordinates.
- Integration of multiple information sources involves using techniques to integrate diverse data sources provided by various sensing devices.
- Linear feature detection and analysis provide analysis of graphic features like roads and buildings and recognition of events taking place in or around them.
- Model-based scene analysis works on the use of geometric models to recognize objects in a scene.
- Natural object recognition develops techniques to recognize naturally occuring objects and events like sky, shadows, and edges.
- Optimization-based recognition involves creating algorithms that predict the functions of objects in order to determine what they represent.
- Partitioning and perceptual organization develop the process by which images are broken down into their basic components for recognition analysis.
- Representation of natural scenes researches several techniques to represent both natural and manmade scenery.
- Stereo analysis involves 3-dimensional model building based on stereo and geometric reasoning techniques.
- Virtual reality (VR) provides technologies that enhance AI interfaces.

SRI maintains a Web site dedicated to AI through their Artificial Intelligence Center (www.ai.sri.com). Further information and a description of current projects involving AI can be found there. Another key research center for work on AI and other aspects of what is known as human-computer interaction is the Massachusetts Institute of Technology Media Lab. This institute had been instrumental in demonstrating that the work of computer science has real world application. A partial listing of projects housed there in late 1998 shows the breadth of thinking taking place around this issue: news in the future, spatial imaging, interactive cinema, gesture and narrative language, sociable media, personal information architecture, tangible media, media that learn, digital life, opera of the future, the future of learning, responsive environments, perceptual intelligence, machine listening, and lifelong kindergarten.

The study of human-computer interaction is one that most researchers feel comfortable approaching in discrete segments, rather than attempting to answer the challenge on as a grand a scale as Cycorp's effort. And while the day when a computer will be able to hold an intelligent conversation with a human may still be many years away, many initiatives are showing promise for easing the interface problem.

Virtual Interface

One of the innovative, though largely experimental, interactive modalities in use for the past five years is the computer-human interface provided by virtual reality (VR) devices. Simply stated, VR is a representation of reality via the interaction of digital output devices. Some of the more interesting developments in this field are highlighted by the CAVE Automatic Virtual Environment project and the Immersa-Desk project, both products of research coming from the Electronic Visualization Laboratory at the University of Illinois at Chicago. Further information on how these immersive, networkable systems work is included in chapter two of this book. Also worth mentioning are the interesting recent improvements to the haptic (something sensed by active touch) devices, which are used in most VR systems today.

Figure 1.2. The Glasstron headset contains two LCD screens that simulate viewing a 52-inch television screen at a distance of 6.5 feet.

Reprinted with permission of Archive Photos/Jeff Christensen/Reuters.

Many systems use the "data glove" which is connected to a computer via a wiring harness that can track movements of the subjects fingers, hand, and sometimes, wrist and elbow. By manipulating the gloved hand, a user has access to the 3-dimensional world being rendered by a speedy microprocessor on a computer's display device (monitor or 3-dimensional goggles of some type). Some data gloves actually contain embedded sensory buttons that can give the operator a sense of feel when objects in the 3-dimensional environment are acted upon or bumped.

Carnegie Mellon University (www.cs.cmu.edu/~msl) researchers have gone a step further down the virtual path by introducing a new type of interface that uses magnetic levitation to physically interact with simulated objects on computer screens. This new device not only allows the users to "reach in" to the 3-dimensional environment and touch objects, but it allows for manipulation in all 3 dimensions as well. Carnegie Mellon has provided these features while at the same time eliminating the cumbersome cable connectors.

Fixed in a desktop-sized cabinet, their system utilizes a light-weight bowl-shaped element that contains six levitation coils surrounded by high-strength magnets. The operator holds a handle that extends from the bowl, which floats in the magnetic levitation field. Because of the gravitational forces present in the strong magnetic field, the operator can feel the force and torque as he reaches into and manipulates objects in the simulated environment. Principal investigator Ralph Hollis of Carnegie Mellon's Robotics Institute and his Ph.D. student, Peter Berkelman, have worked five years to develop this haptic interface.

Spoken Language Systems and the Conversational Interface

The ultimate human-machine interface in most people's view would be a conversation in which the computer would be capable of interpreting what the user wanted and then be able to respond in synthesized speech answers that made sense in human terms. A conversational interface would allow for the creation, access, and management of information and to solve problems. Since 1989, the Spoken Language Systems (SLS) Group at MIT's Laboratory of Computer Science (www.sls.lcs.mit.edu/sls) has researched how to make this interface work.

Speech recognition software is much better now that processors have the power to calculate the necessary advanced algorithms in shorter time periods. Also, current interactive voice systems that call for a synthesized computer voice-response in commercial applications have also improved. But as good as they are, these systems still require the human to address the machine in narrowly defined, immutable steps. The MIT group is working on procedures that go well beyond these capabilities. In SLS, the interaction will be spontaneous with the end result being reached in far fewer steps. To date, the SLS researchers have launched several prototypical systems that approach the lofty goal they have set for themselves.

A breakthrough came in 1994 when the group developed a conversational platform that they have named GALAXY. The work recently advanced further when this software engine was combined with the power of several high-performance servers dubbed SUMMIT (which handles speech recognition); TINA (which handles understanding languages); and GENESIS (the language generation server). The sequence for their Jupiter Program (a phone-based system that will allow users to ask the computer questions about weather conditions in hundreds of cities throughout the world) follows this outline:

- SUMMIT retrieves the query from the user and converts the spoken sentence into text (voice recognition).
- TINA then parses the text that has been generated into a grammatical structure that contains the terms appropriate for a query to the Jupiter Database of weather conditions that have been gleaned from Internet sources (language understanding).
- GENESIS next uses the semantic frame generated by TINA, using the basic terms to build a Structured Query Language (SQL) query for the database (language generation).
- Jupiter immediately executes the SQL query and assembles the requested information from Jupiter data (information retrieval).
- TINA and GENESIS work together to convert the SQL result into a natural language sentence (language generation).
- Jupiter uses a speech synthesizer to answer with the generated sentence from the step above. It can also display the answer for those who cannot hear. And depending on how they first identified themselves, users will be answered in either English, Italian, or Japanese.

The Jupiter service started counting inquiries in May of 1997 and has fielded over 100,000 calls with a word accuracy of nearly 89% for those who are not familiar with the system. Jupiter does much better when experienced users access the system, getting over 98% of the words correct in those instances. Jupiter knows over 2,000 words. To try the system, use the toll-free number from North America, (888) 573-8255, or the long distance call from anywhere else, (617) 258-0300.

Besides Jupiter, the MIT SLS Group is also responsible for GALAXY-based conversations about airline scheduling (PEGASUS), Cambridge city locations (VOYAGER), Boston area restaurants (DINEX), automobile classified ads (WHEELS), and selected Web-based information (WebGALAXY).

NEW PROGRAMMING MODELS

There is a growing movement among computer science professionals to revisit the processes by which software is presently produced. This is especially true when these professionals analyze the development and implementation of commercial programs licensed for the mass market. For some time, critics have lamented the fact that applications are released before they are fully tested, programs are bloated and unstable, and that the "bottom line" in code writing is really dollar signs and not elegant implementation of a programmer's vision. Some programmers have posited that a return to the good old days of cooperative code-sharing, now represented by the open source movement, would go a long way toward correcting programming problems that make intelligent machines harder to use than they need to be. Others advocate a new approach to programming basics.

Rethinking the Sequential Software Development Model

Dr. Lynn Andrea Stein (www.ai.mit.edu/projects/cs101) teaches a class to incoming computer science students at the Massachusetts Institute of Technology (MIT) called "Radically Rethinking Computer Science 101." In it, this programming groundbreaker hopes to get her message across regarding the new realities of computing and how software functions have been radically altered in the past decade. The most important difference is in computations that involve concurrent interactions with users, networks, and environments like the World Wide Web. In Stein's opinion, it takes new approaches in thinking to design these types of computational systems. As she noted in an interview she gave to *Face To Face*, an online Java programming magazine produced by Sun Microsystems,

> Today's programmer can never be ignorant of the context in which a program will run. There are always many things going on, inside your program and around it. Today you have multiprocessors, multiple-issue machines, multithreaded programs, and you have users who operate concurrently—while the CPU is processing the results of my mouse click, I'm moving my mouse and starting another task. For example, a modern word processor works by activating several things simultaneously to organize the information on the page, doing what the old computation model did sequentially. It's checking my spelling, correcting my gram-

mar, and inserting capitals at the beginning of sentences automatically while I'm writing, all at the same time.

The way that you evaluate this kind of system is by judging its ongoing interactions over time. You don't wait for the program to come to produce its result and stop—if the Web, or a robot stops, it's broken! At the same time, we need to understand what the entities are, how they interact, and how we're going to design each one. That's where recursive decomposition comes in—programmers actually need to know how to decompose these types of programs into concurrent tasks.

That means they need to know how to think concurrently. The questions become: What are the services my system provides? Who are the entities that make up the community that is my system? How do these entities interact to provide those services? What is each entity made of, for example, how does it provide its own services within this community (Byous 1998)?

The commonly used system of programming ignores the concurrency described by Dr. Stein. She notes that it is still based on the "stored program" ideas of mathematician John von Neumann. In his classic conceptualization, free-standing objects exist independently of other applications, and they are tied together in a series of sequencing steps. According to Stein, "That's not reality." Today, the computer works in a simultaneous mode of interdependencies. Sequential steps, while enabling computational theory to evolve, totally ignore what the processor might be doing while it moves through the programmer's code (signaling peripheral processors, receiving user input, etc.). This decision to ignore reality, or what is known as sequential abstraction, has led to bloated code and inelegant software. Changing development patterns in the software writing community will not be a simple task. But, there are already movements afoot to produce code like that advocated by Stein.

Easier Computing

Transparent (meaning easy to use or intuitive) and simple are essential elements in the new vision of computers and connectivity that are far from given. Some progress certainly has been made in creating hardware and software that is more user friendly over the preceding decades, but there is still room for developing devices and software that are easier to use, more streamlined, and as bug-free as users have a right to expect.

Robin Gunier pointed out in *Computer Weekly*, a British journal covering the industry, the ideal would be for

a business and domestic world where the immense power and value of computing devices would become available to people who neither had, nor cared to acquire, the arcane skills required today.

Just as it is now possible—and usual—to drive a car without knowing the first thing about the internal combustion engine or the automatic gear box, and for that no longer to be the slightest disadvantage, so totally nontechnical people will be able to get as much value from computing devices as do the nerds. (Gunier 1998)

Bill Gates, CEO of Microsoft Corporation, made the same point in his keynote address in front of the COMDEX attendees at the fall 1998 event in Las Vegas, Nevada. Gates displayed some of the latest advances in hardware power along with new software-based enhancements that are designed to make computing more natural. "These include 3-dimensional, real-time effects displayed on a Silicon Graphics workstation, enhanced type display that will make electronic books feasible, and wireless input devices that free the user to walk away from the machine, but remain connected. As PCs become more powerful, we must work to keep them simple" (Gates 1998). Gates goes on to say that the goal of Microsoft's Office 2000 and Windows 2000 (the latter being the newest operating system in the Windows family) is the creation of a more stable system for the professional and consumer user.

Microsoft has defined the industry and the personal computer interface through its domination of the market place with the products MS-DOS, Windows, Windows 95, Windows 98, and Windows NT. Each iteration of the Microsoft operating system has become more intuitive and easier to use as the graphical user interface (GUI) has improved and new features that integrate common tasks have been added.

The Potential of Open Source

As an alternative to the model of large corporations dictating the way devices are operated (selling large, feature-rich applications that are continually enhanced through expensive upgrades), many programmers advocate opening up the software development process so that simpler computing environments can be achieved. The idea behind "free software" or "open source" is a simple one: make the underlying code for programs (applications or operating systems) free for the taking. Everyone and anyone who wants to can then download the basic software to run it, or better yet, modify and improve it. The program evolves and develops, usually becoming more stable and more useful in a very short time. For instance, it is estimated that a minimum of 1,000 programmers were, or are, involved in developing Linux and applications that run on top of it. And that effort has brought about a significant change in the possibilities of enterprise computing.

Linux

Linux is the Unix-like program that was begun by Linus Torvalds at the beginning of the 1990s. His student project became a collaborative and freely cooperative enterprise when he sent a message to other programmers via a Usenet newsgroup on July 3, 1991.

Unix, Minix, Posix, and the popular variation, Linux were all attempts to create time-sharing operating systems for computers connected over networks. Torvalds' plan was to develop his variant of Unix under the GNU Project's General Public License standard. The GNU idea had developed some eight years earlier when Richard Stallman first tried to design his own version of a freely distributed Unix alternative. Linux is the system that succeeded like no other. It supports 32-bit or 64-bit multitasking, allows hundreds of users to access a single computer at one time via its very fast TCP/IP (transmission control protocol/Internet protocol) drivers, and, unlike Windows NT, works in a fully protected mode.

The buzz at the 1999 fall COMDEX show *was* Linux. Even though it had been a favorite of programmers and a large number of Web server technicians for many years, the reliability problems of other network-centric operating systems (especially Windows NT), catalyzed the efforts of major vendors like Oracle, Informix, Netscape, and Intel to back Linux initiatives. The latter two companies, along with others, have invested in Red Hat Software Inc., which is the main distributor of the free Unix clone and collects revenue for bundling tech support, documentation, and applications with the operating system (OS) in order to boost Linux acceptance throughout commercial enterprise networks.

At a time when some were calling the open source movement "programming anarchy," this investment by some of the biggest players in the industry has given it substantial credibility. The PERL programming language, the Apache Web Server, and Sendmail were some of the historical high points of the free code movement, but the Linux product could become the ultimate grassroots success story. Predictions that it will be widely implemented as an alternative to Microsoft's expensive solution are being seen more and more in the influential industry publications (Torvalds 1991).

Netscape Opens Its Vault

In a surprise development in early 1998, Netscape Communications made the code for its Navigator Web browser product freely available for downloading from its company servers. By making its core code for

Navigator 5.0 available to the thousands of programmers via the Internet (over 4,500 downloaded it the first day), Netscape signaled that the browser war against Microsoft Internet Explorer was entering a new phase. The expectation was that by opening the development process to the Internet community, subsequent versions of their product would contain exciting new features and be more stable than what could be accomplished in-house. The company announced that within the first 48 hours, they were able to modify their working product based on feedback they had received from various programmers from around the world.

Alan Shutko, a developer, was quoted in an article on CNET's News.com regarding the importance of the open source movement and Netscape's initiative in this area:

> Originally, the World Wide Web was almost entirely free software. This triggered lots of good ideas, and was the reason that it took off. But recently, Microsoft and Netscape have been locked in a browser war, which has only given us buggy browsers. Now there will be less of a stress on Netscape employees and it will be easier to advance the technology because there will be so many people helping.
>
> It'll also be quite valuable to read the source, simply to see the issues involved in this kind of application. The more source that's out there to read, the better programmers can become. It's like literature: You will have a hard time being a good writer unless you've read a lot of books. But we expect programmers to be good programmers without seeing many different approaches to systems and problems. Netscape's source release will help that (Kornblum 1998).

SUPERCOMPUTING

The Alex Informatics AVX3, the Cray Research C90, the Cray Origin2000, the Convex C4600, the Convex Exemplar SPP1000/XA, the Hitachi S-3800/480, the IBM SP2, the Intel Paragon XP/S MP, the SGI Origin 2000, the nCUBE, the nCUBE2, the Pyramid Reliant RM 1000, the NEC SX-4/32, the SGI Power Challenge XL, the Tera MTA 32, and the Thinking Machines CM-5—these exotic designations are all names for some of the worlds most powerful supercomputers, the epitome of high-performance computing in the scientific research community.

In the 1980s, the mainframe computing environments with their customized vector processors defined the supercomputer. Used for resource-intensive number-crunching functions, these large connected cabinets often still filled the space in dedicated rooms. But as the 1990s dawned, the same power that had once been available only in these rarified domains was now starting to show up in the new microcomputer architectures and even in personal computer desktop solutions. The power of Moore's Law was taking hold. Instead of using the vector processor as the CPU for the supercomputer, the use of parallel processing systems that could utilize more cost-effective, off-the-shelf chips became a standard for a new generation of machines.

Power, in the realm of supercomputing, is measured by the number of floating-point operations that are completed in a second (flops). The benchmark goal for years was one trillion floating-point operations per second or one teraflops. The machine that first achieved this mark was an Intel system that was set up to use 7,264 Pentium Pro chips in a parallel infrastructure. Since the first teraflops performance in 1996, speeds have continued to increase, through a combination of reconfigured software, more powerful microprocessors, and improved architectures and algorithms. IBM was in the vanguard of this development with their family of machines built on the Deep Thought architecture. Many readers may recall that this machine was programmed to play the highly publicized chess matches against world grand master Garry Kasparov. Eventually, an improved computer, Deep Blue, was able to defeat Kasparov in 1997. Other typical super-computer applications include number crunching for computational scientists and modeling large systems like weather or body functions.

By 1999, one of the top supercomputers was the latest in IBM's line, the Blue Pacifier. It boasts 1.2 teraflops of sustained performance with a measured peak output of 3.9 teraflops. This power was achieved through

Figure 1.3. The "Teraflop" Supercomputer at Sandia National Laboratories.

Sandia photo by Randy Montoya. Source: Sandia National Laboratories, Albuquerque, NM.

a fixed hypercluster array of 5,856 Power PC 604 processors with a random access memory capacity of 2.6 terabytes. The unit, which stands seven feet high, occupies 8,000 square feet of floor space, requires 5 miles of power line and over 50 miles of copper and optical cable, was produced at a cost of $94 million.

Then, in February of 1999, a start-up company in Salt Lake City made a surprise announcement that altered the supercomputing environment once again. According to the press release from Star Bridge Systems, Inc. regarding their new HAL-4rW1 Hypercomputer System,

> These systems are massively parallel, reconfigurable, third-order programmable, ultra-tightly-coupled, fully linearly-scaleable, evolvable, asymmetrical multiprocessors. They are plug-compatible supercomputers that surpass conventional supercomputers in features and performance, at a far more attractive price. Unlike any other supercomputer available, they perform a wide range of computationally intensive tasks in real time in an extraordinarily small amount of hardware. Target market: All supercomputer applications and many applications beyond the capacity of conventional supercomputers (Starbridge 1999).

When compared to the IBM standard, they appear to be remarkable machines. After working on his hypercomputer concept for 15 years in his Sandy, Utah, home, Kent Gilson, Star Bridge Systems Chief Technical Officer, has designed a system that he says is about the size of a personal computer, is capable of running multiple operating systems, and can actually reprogram its 100 billion circuits on its own. This machine, which is plugged into a normal wall outlet has shown speeds of up to 12.8 teraflops.

It gets its ultra fast performance boost from a radical new architecture that is called "reconfigurable computing." The HAL system includes about 100 billion circuits which can be reprogrammed on the fly, not through human intervention, but by software and the machine itself. This means that a circuit optimized for a particular task can be instantly rerouted to perform another function in a new optimal state. This reprogramming can take place thousands of times per second. The makers foresee the day in the very near future when all information appliances will contain this type of chip.

References

Air Force. 1998. "AFRL-Rome Awards $2.1 Million Contract to Syracuse University." http://www.if.afrl.af.mil/div/IFO/IFOI/IFOIPA/press_history/pr-98/pr-98-48.html. (2 February 1999).

Byous, Jon. 1998. "Dr. Lynn Andrea Stein: Radically Rethinking Computer Science 101." *Face to Face* (java.sun.com). http://www.javasoft.com/features/1998/06/lynnstein.html. (10 April 1999).

DeTar, Jim. 1998. "National Semiconductor to Put PC on a Chip." *Semiconductors International*, 7 July 1998. Highlands Ranch, CO: Cahners Publishing Company.

Gates, Bill. 1998. Keynote speech at COMDEX, Las Vegas, Nevada, November 15, 1998.

Georgia Tech Research News. 1998. Press Release: "A New Computing Paradigm: Chaos-based System That 'Evolves.' Answers May Be Alternative to Current Computing." *Georgia Tech Research News*, 8 September 1998.

Gunier, Robin. 1998. "Make Them Work for Their Money." http://www.computerweekly.co.uk. (1 October 1998).

Kornblum, Janet. "Developers Flock to Netscape Code." CNET. http://www.news.cnet.com. (3 April 1998).

Merkle, Ralph. 1998. http://www.Zyvex.com/nano/. Zyvex L.L.C. (8 September 1999).

OE Reports. 1999. "Self-Assembling Plastics Promise Useful Optical Devices." *OE Reports*, Number 183, March 1999. www.spie.org. (16 January 1999).

Science Daily. 1998. "Carbon Nanotube Discoveries at Georgia Institute of Technology." *Science Daily*, 18 June 1998.

Starbridge. 1999. Starbridge Systems. http://www.starbridgesystems.com. (5 February 1999).

Texas Instruments. 1999. "Texas Instruments History." http://www.ti.com/corp/docs/history/firstic.htm. (25 June 1999).

Torvalds, Linus Benedict. Newsgroup message "comp.os.minix," 3 Jul 1991. http://www.li.org/history/index.shtml. (15 January 1999).

CHAPTER TWO
Building the Next Generation Internet— Research and Developments

T he faster computers, better code, and enhanced storage break-throughs that are the subject of the previous chapter have provided the basic building blocks for the foundation of a new computing paradigm. This developing vision seeks to replace the current model of the individual, autonomous, desktop PC with a more fully connected "distributed computing architecture." This environment supports the use of applications and data across a widely distributed network of different types of computers and operating systems.

As greater connectivity to the Internet has demonstrated, the computer user's ability to communicate, collaborate, and interact with others across the planet has increased at an exponential rate. And with the proliferation of hundreds of smart devices (appliances containing embedded chips), this is no longer just the computers domain. Many researchers are working to make those connections even more powerful, more ubiquitous, and a lot more human. To achieve the promise of the type of distributed computing infrastructure that can support the applications that the scientific community, business, media, and government interests will require, the Next Generation Internet (NGI) initiative is underway.

The federal government, private industry, and scores of universities are supporting work on parts of a broadly defined vision, which has the potential to alter reality beyond anything yet experienced since the onset of the Information Age. This chapter looks at two of the most important research consortia now developing the digital tools that will make this happen. Funded by the National Science Foundation, the National Computational Science Alliance and the National Partnership for Advanced Computational Infrastructure are administering parallel and complimentary programs across the United States. What follows is a brief overview of the objectives and some of the more important software, hardware, and interface initiatives supported by the architects of the new computational grid.

WHAT IS THE VISION?

> A computational infrastructure. Cyberspace on steroids. A virtual machine. Imagine taking apart various desktop computers and souping up each of the components—memory, CPU, graphics devices—so that they are thousands of times faster and orders of magnitude more sophisticated than they are now. When you connect them via a network that shoots data across the country at speeds approaching that of light.... What you end up with is no longer a computer but a portal to a flexible and powerful computing environment. That's the Grid (NCSA 1999).

This goes beyond and incorporates the effort to build the NGI, also known as Internet2 (www.ngi.gov). That initiative has the support of the federal government, which authorized $225 million toward its development through 2000. The work is being done by a partnership of corporations, research centers, and universities that have joined to build a new broadband Internet for the delivery of integrated voice, video, and data between higher education institutions across North America. The goal of NGI is to provide for a research communication infrastructure that will provide standards for broadband (high capacity data delivery) nets in the future. Those building the grid have accumulated a great deal of experience to help them achieve the goal.

In 1986, funding by the National Science Foundation (NSF) opened the Supercomputing Centers Program with the intention of developing a network of research institutions that could share experience, resources, and knowledge with each other and the world. The first supercomputing centers were located at the Cornell Theory Center (CTC), the National Center for Supercomputing Applications (NCSA) at the University of Illinois at Urbana-Champaign, the Pittsburgh Supercomputing Center (PSC), the San Diego Supercomputer Center (SDSC), and the John von Neumann Center at Stanford University. The goal then was to advance computing power and efficacy and to maintain the United States' leadership in science and technology. The legacy of the first decade of that program is a remarkably productive one, featuring new applications in high-performance computing, visualization, and desktop software. Perhaps the most famous product to emerge from this effort in the 1990s is the graphical Web browser NCSA Mosaic. Even though other Internet protocols had set the stage for interaction and collaboration between scientists and scholars using remote-based applications, the Mosaic graphical display opened a door to the next level of interaction.

Beyond this crucial effect on the research community, that one application was the basis for a popularization and redefinition of computers and computation in the minds of most Americans. A remarkable and incredibly accelerated transformation of society began when users started to "surf the Net," affecting everything from telecommunications applications and infrastructures to business productivity and the means of exchange. Consider the changes that have taken place throughout society since graduate student Mark Andreessen left NCSA and the University of Illinois in 1995 with the software code he had helped develop. He joined the start-up company Netscape Communications of Mountain View, California, which further popularized the use of the Web with its ubiquitous Netscape Navigator application. This development and other commercial applications will be covered in more detail in chapter 3, but it is important to understand that much of the next wave in computing machines and applications has a genesis in the academic and private research facilities.

THE NATIONAL SCIENCE FOUNDATION LINK

To expand on the groundbreaking work that the supercomputing centers have made since 1985 and to leverage the power available at the countless separate computer research centers throughout the United States, the National Science Foundation set aside nearly $21 million in 1997 to establish the next generation of network collaboration. In making this decision, the directorate relied heavily on several reports that had analyzed the Supercomputer Centers Program and that had made recommendations for future computational paradigms. The most relevant one is the *Report of the Task Force on the Future of the NSF Supercomputer Centers Program*, chaired by Dr. Edward Hayes, and published in 1995. The report discussed the history of, and the rationale for, NSF support of a high-performance computational infrastructure that would benefit the science and engineering research communities. In addition, two other earlier reports, *From Desktop to Teraflop: Exploiting the U.S. Lead in*

High Performance Computing, chaired by Dr. Lewis Branscomb, and *Evolving the High Performance Computing and Communications Initiative to Support the Nation's Information Infrastructure*, chaired by Drs. Frederick Brooks and Ivan Sutherland, provided additional background and supported the need for a high-speed network that would facilitate research among science communities (NSF 1995).

The program that took shape as a result is known as Partnerships for Advanced Computational Infrastructure (PACI). In its request for proposals, NSF wanted to fund entities that could meet the following objectives:

- Provide access to a wide range of advanced computers, data storage systems, and experimental machine architectures.
- Supply "enabling technologies" through the development of software tools for parallel computation, and the creation of software that would facilitate use of widely distributed and architecturally diverse machines and sources of data.
- Involve scientists using high-performance applications so that their discipline-specific codes and software infrastructures could be made available to the program, as well as to researchers in other areas.
- Build awareness and understanding of how to use high performance computing and communications resources, and broaden the base of participation in these areas.

On October 1, 1997, NSF chose two parallel programs that would constitute the Partnerships for Advanced Computational Infrastructure. Those programs are the National Computational Science Alliance and the National Partnership for Advanced Computational Infrastructure. In both of these consortia, the overriding objective is to put a structure under the diverse pieces of the research milieu: to develop a way for scientists to interact among their own and other interested communities through the implementation of a broadband network that will leverage computing power to a very high degree. Later in this chapter, we will be looking at the initiatives that fall under the umbrella of National Partnership for Advanced Computational Infrastructure headquartered in San Diego. But we begin the survey by looking at the very diverse body of work being organized under the direction of the University of Illinois Supercomputing Center, NCSA.

THE SUPERCOMPUTING ALLIANCE COMPUTATIONAL GRID

The National Computational Science Alliance, whose leading-edge site is the National Center for Supercomputing Applications at the University of Illinois, Urbana-Champagne (www.ncsa.uiuc.edu/alliance) is "a partnership among computational scientists, computer scientists, and professionals in education, outreach, and training at more than 50 U.S. universities and research institutions working to prototype the computational and information infrastructure of the next century" (NCSA 1998). The Alliance, as it is known, is charged with developing an advanced communications infrastructure: the National Technology Grid (the Grid). Over the tenure (1986–1997) of the first supercomputing program there was an exponentially increasing demand among researchers for high-performance computing access. Since it is expected that scientists will pursue increasingly complex problems, administrators are projecting that the need for computer cycles will reach 10 teraflops within the next five years. (That is 10,000,000,000,000 floating-point operations per second. Teraflops is the standard unit for measuring high-performance computing.) Currently, more than a thousand projects are being computed at NSF supercomputer centers. The problem is that simply building larger, faster computers will not meet the demand for resources. "What is required is a distributed computing grid populated with powerful computers that are individually scalable and that, when necessary, can be assembled into metacomputers" (NCSA 1998). This system, when brought online, will leverage the power of coupled parallel computers, workstations, large databases, and other resources like virtual reality devices connected via networks.

The work of the Alliance is not only deployment of the NGI, but the basis for a new method of collaborative computing and knowledge development. Modeled after the national electric grid, which transformed society in the last century by making electrical power available to citizens on demand, the hardware and software that the Alliance is developing will eventually give every user access to a privately funded high-performance, communications network of staggering computational power. Use of this computational grid will be as normal and as widespread as use of the PC is today.

To better understand how this ambitious goal might become a reality, it is best to survey some of the current research efforts taking place under the broadly defined Alliance umbrella. Member institutions are focusing their research and discovery efforts under four basic divisions.

- *Applications* work is initially concentrated in data- and computer-intensive areas that include chemical engineering, cosmology, environmental hydrology, molecular biology, nanotechnology, and scientific instrumentation. Collaboration software, data mining, and visualization are also important tools that will eventually make the Grid technologies useful for broader societal needs beyond science.
- *Programming Tools* are of crucial importance because a flexible user framework is an important condition for hiding some of the complexity involved in the underlying infrastructure. The goal is to have the technology adapt to the user, not force the user to learn new tools each time the system is accessed.
- *Services* is the division of the Alliance effort that will attempt to create a middleware software solution for accessing the diverse hardware that will be the basis of the Grid computing environment. One analogy is that this middleware is like the operating system on a PC. The services will provide security, authentication, resource scheduling, and quality of service functions.
- *Physical Resources* are the high-performance computing facilities, including computers, visualization environments, mass storage devices, and networks.

The organizers do not expect that research efforts will always fit neatly into the above categories, as advances in design, deployment, and the underlying technology always overlap. But results from Alliance-associated initiatives will be evaluated and integrated as per that design. Some of the following research programs were underway before the Alliance was formed. Their successes will be leveraged into the ultimate goal of creating the Grid.

The Alliance's job is not an easy one. It must mesh diverse initiatives on myriad architecture and software platforms taking place in research centers widely distributed around the country. At first glance, it would appear that each lab is working independently on pieces of a puzzle whose outcome is only generally defined as "more powerful computation." The Alliance has implemented three Enabling Technologies Teams (Parallel Computing, Distributed Computing, and Data/Collaboration) to facilitate the corralling of these powerful, yet diverse efforts. It fully understands that scientists who have spent years developing more powerful code to answer ever more complex questions more efficiently on their installed base of computers are loathe to divert those efforts to reprogram in order to operate on another platform. But if these scientists are to take advantage of the developing architectures, especially the emerging schemes that link divergent hardware running different operating systems in parallel processing operations, they will have to migrate. The result will be that larger databases can be accessed, more quality displays of instrument images can be seen, and more realistic simulations can be run.

Parallel Computing

A key objective of the Alliance's Enabling Technologies Parallel Computing Team is to build a toolkit that will save the scientists from having to rewrite their most effective applications so that they can still take advantage of the scalable performance of parallel architectures that can use many microprocessors at one time. This software toolkit will contain applications that help to modify and move proven applications so that they can run on distributed shared-memory (DSM) systems that allow computers on a network to use the same memory pool, or on the new commodity (off-the-shelf workstations) cluster systems.

> The toolkit will enable researchers to readily use the architecture best suited for a given job. In the Parallel Computing Team's toolkit, for instance, will be a set of portable programming languages and compilers. Compilers translate codes from human-readable programming languages into the instructions understood by computers. A program written in a portable language can be easily translated to execute on different computers. Among the first projects to be tackled by the team is the development of new compiler capabilities for high-performance Fortran (HPF) and high-performance C++ (HPC++) running on DSM systems. Ultimately these tools will enable scientists and engineers to construct codes that port to any architecture in the Alliance. The toolkit will also include libraries—encapsulated code that can be incorporated into applications to carry out complex communications and mathematical functions frequently called for in high-performance computing. The Parallel Computing Team will collaborate with Application Technologies Teams to optimize libraries that are applied in many settings. Linear solvers, for instance, are at the core of thousands of scientific applications ranging from molecular biology to nanomaterials (NCSA 1998).

Distributed Computing

The Alliance's Enabling Technologies Distributed Computing Team is leading the effort to exploit the substantial advances of multiple vendors and researchers via the distributed computing environment of the National Technology Grid. While self-contained desktop computing on a PC has the advantage of using integrated storage, processing, and graphics systems due to its single vendor source, a distributed computing environment can work much the same way if the distributed computing team can reach its goal of creating a "virtual metacomputer." This will happen if the team's integrated software system can implement an interface between all of the various processing systems that can be brought to bear along the connected network crossing thousands of miles.

Similar to systems software on desktop computers, the Grid's operating environment manages the components of the metacomputer. Think of this environment, or middleware, as providing a collection of services: mechanisms for communicating, accessing code and data, measuring performance, guaranteeing quality of service, and ensuring privacy and security. Scheduling—the software process that reserves both the equipment and the network connections needed to run applications—is one of the most challenging services to render. The global scheduler must be flexible so that it can accommodate the scientist or engineer who needs hundreds of processors for a run lasting hours or weeks as well as the radio astronomer who must interactively access remote telescopes to execute large sets of real-time observations (NCSA 1998).

In regard to the scheduler problem, the Enabling Technologies Distributed Computing Team is looking to the groundbreaking work done in Globus under the auspices of Argonne National Laboratory and the University of Southern California's Information Sciences Institute. By integrating this software with the University of Wisconsin Condor and NCSA Symera software, the Alliance will enable both local and global linking of resources, providing a common computing framework across the nation. Further information about Globus, Condor, and Symera, as well as details about the critical initiatives, visual supercomputing and tele-immersion, is provided later in this chapter.

Data and Collaboration

The speed at which data moves across the network that will become the National Technology Grid will be a key to judging how successfully scientists, researchers, and eventually all users will be able to interact. Collaboration in real time, regardless of the speed of currently available microprocessors, will not be feasible if the input/output (I/O) bottleneck is not broken. Currently, scientists require access to data sets that are up to a terabyte, or 1024 gigabytes (10^{12}), in size and that may be located in mass storage systems (tapes and disks) hundreds of miles away on the network.

High-performance computers can operate at speeds exceeding a trillion operations per second, but I/O operations run closer to 10 million bytes per second on a state-of-the-art disk. By making available a toolkit that developers can use to create and control related file systems, the Enabling Technologies Data and Collaboration Team will make possible large speed increases in I/O, bringing these operations more in line with processor performance. This team is also deploying information management methods to organize, characterize, and access data efficiently so that it can be shared more easily. Various user interfaces will enable researchers to enter their own metadata—labels that describe the context and structure of a file's data—then index and extract information based on these characteristics (NCSA 1998).

The third objective of the Data and Collaboration Team is, as their name implies, enhancing collaboration. One emphasis is high-modality, immersive data exploration and collaboration, so that researchers can manipulate and explore data simultaneously via virtual reality environments. Powerful tools such as Virtual Director are enabling researchers to steer, edit, and record their navigation through a large data set. Another emphasis is on ubiquitous desktop collaboration. The infrastructure is provided by NCSA Habanero, a Java-based framework for synchronous (realtime) and asynchronous (delayed) collaboration, together with DistView from the University of Michigan and TANGO from Syracuse University. Incorporated into the infrastructure will be repositories, meeting spaces, database services, and even desktop virtual reality.

Networking Initiatives to Build the Grid

The privatization and commercialization of the original Internet has led to frequent periods of heavy congestion. These data traffic jams have negatively affected the university research community's ability to communicate and collaborate. In response to this challenge, research centers have been working on several initiatives to create a new network that can withstand the ever-increasing traffic. Among those discussed in this section are Internet2, The Abilene Network, vBNS, MREN, STAR TAP, and CANARIE. The most ambitious of these is Internet2 (www.internet2.edu).

Also known as Next Generation Internet (NGI), the Internet2 project started in October 1996 when 34 universities decided to provide resources and facilities so that their communities, along with government and industry partners, might accelerate the next stage of Internet development in academia. Today, over 130 universities and many corporations are partners in the project. By establishing applications to fully exploit the capabilities of broadband networks, media integration, and interactivity, it is expected that new national research objectives, distance education deployment, and lifelong learning opportunities can be realized. Eventually, new network services will be accessible to all levels of the national education community, and then to general users across the globe.

Georgetown University

One of the Internet2 universities, Georgetown University (www.georgetown.edu/research/i2/) reported on one of its first attempts to use the high-capacity system at the World Congress on Information Technology in June 1998. Their collaborative Telemedicine project with George Mason University (GMU) used Georgetown's Imaging Science and Information Systems (ISIS) to send large sets of magnetic resonance imaging (MRI) brain scans to GMU. Once there, the sets were reconstructed as 3-dimensional images in real time. At the same time this massive data transfer was occurring, they also used the network as a videoconferencing medium, connecting doctors and other participants at each site.

The Abilene Network

The Abilene Network is another broadband network project. This one is operated under the auspices of the University Corporation for Advanced Internet Development (UCAID) (www.ucaid.edu/abilene). It is supported by over 175 member organizations, many of which are the same universities now associated with the Internet2 program. UCAID also includes corporate partners like Cisco Systems, Qwest Communications, and Nortel (Northern Telecom), with whom UCAID is deploying an experimental research network, which will initially use OC-48 (2.4 gigabits per second) "backbone," or main data carrier, links. Plans are to develop links that move at the rate of OC-192 (9.6 gigabits per second) and greater. The intent is to provide an alternative and ancillary backbone to the very high performance Backbone Network Service (vBNS), which was instituted by MCI in 1995 under a five year agreement with the NSF. This backbone will be there as an interconnection between the regional networking aggregation points (large point of presence facilities containing routing equipment for the network and known as gigaPoPs) of the Internet2 universities, so that the universities can develop new software applications faster and more broadly.

The vBNS

The vBNS (www.vbns.net) is an MCI nationwide network operating at a speed of OC-12 (622 megabits per second), using Asynchronous Transfer Mode (ATM) and Synchronous Optical Network (SONET) technologies. The vBNS is the product of a five-year cooperative agreement between MCI and the NSF to provide a high-bandwidth network for research applications. Over 50 of the Internet2 university gigaPoPs get their connection via vBNS, and two of the nation's supercomputing centers are connected. In addition, MCI provides a second "testbed" network to experiment with interoperability and emerging communications systems.

Metropolitan Research and Education Network

Metropolitan Research and Education Network (MREN) (www.mren.org) is part of the UCAID group and is helping to define Internet2 by creating an advanced form of the gigaPoP. Basic research in this area is focused on the challenge of building networks that support multiple classes of traffic while still optimizing network resource utilization. Typically, Asynchronous Transfer Mode (ATM) technology has been used as the carrier of choice for supporting multiple applications over networks, but MREN has undertaken projects that test how new scientific applications will work over emerging network protocols. It is also engaged in research to define the way regional research and educational networks will eventually interface with the high-speed pilot networks like Abilene and vBNS.

Science, Technology, and Research Transit Access Point

Science, Technology, and Research Transit Access Point (STAR TAP) (www.startap.net) "is a persistent infrastructure to facilitate the long-term interconnection and interoperability of advanced international networking in support of applications, performance measuring, and technology evaluations" (EVL 1998). Under the leadership of the UIC Electronic Visualization Laboratory, this initiative acts as a large switch, connecting MREN and the vBNS to international networks like the Canadian Network for the Advancement of Research, Industry, and Education (CANARIE); the Singapore Research and Education Network (SingaREN); Taiwan's TAnet; Russia's MirNET; and the Asian Pacific Advanced Network Consortium (APAN). In short order, connections to many other research nets will take place.

Alliance Software Development to Facilitate the Grid

One of the most important areas of inquiry within the Alliance structure is in the creation and modification of software that will lead to better collaboration between sites via an almost transparent system of new tools. Hundreds of projects are underway to make programs more responsive, so that work in diverse areas of study will be easier. A few representative projects illustrate the breadth of the work taking place to make the Grid a reality. Emerge, NCSA Habanero, Lattice Crystallographic Tool, Hierarchical Data Format, Symera, GLOBUS, High Performance Virtual Machines, TANGO Interactive, and Condor initiatives are explained in this section. We begin, however, with the NCSA Biology Workbench program.

The Biology WorkBench

The Biology WorkBench (biology.ncsa.uiuc.edu) is a computer interface developed at NCSA that allows a researcher to use a typical Web browser to readily access bioinformatics (biology data sets). The Biology WorkBench application has successfully integrated standard protein and nucleic acid sequence databases, and a wide variety of sequence analysis programs into a simpler one-step interface. Computational scientists (and teachers and students in K-12) have the luxury of performing database searches that produce analyzed results that do not have to consider incompatible file formats. WorkBench is a model for the way computational resources can be shared over the Web, running as it does on NCSA servers.

Emerge

Emerge (emerge.ncsa.uiuc.edu) is the NCSA initiative that looks at one of the critical problems of storing and retrieving huge quantities of data over distributed systems like the Internet. The goal of Emerge is to build a new search infrastructure that can address the issues of scale and commonality over a high-capacity wide-area network. In the past, indexing and retrieval was only performed on relatively small, manageable collections of data. Software that controlled storage and the software that retrieved the right data were both housed under the same administrative structure. This helped to assure a uniformity. With remote distribution of data across the Internet, it is not easy to assure that search, storage, and retrieval functions are compatible. Emerge strives to standardize search clients and services interoperability requirements using emerging tools.

NCSA Habanero

NCSA Habanero (www.ncsa.uiuc.edu/SDG/Software/Habanero) is a framework for remote collaboration that uses a defined set of applications. Using the Habanero Application Programming Interface (API), which is based on Sun Microsystems' Java Virtual Machine (an application that allows Java programming to work on various platforms), a researcher or student can create and work in shared applications from locations connected anywhere over the Internet. This is a new way to build groupware applications faster. The framework provides a structure for developers to create or convert existing applications into collaborative ones.

The Lattice Crystallographic Tool

The Lattice Crystallographic Tool, which was originally designed by Tai Y. Fu at the University of British Colombia, is used to skeletally view model protein data bank files of molecules in three dimensions. Within this group of applications is the Whiteboard for displaying images; Audio Chat, a Web phone that allows participants to verbally communicate with other collaborators online; and Chat, a common-looking chat program that does not use the normal protocol, Internet Relay Chat (IRC) for real-time messaging. Instead Chat uses the Habanero object sharing mechanism. The Voting Tool is included so that collaborators have the ability to vote on an issue during a collaborative session, and the Visible Human provides access to the visible human database from the National Institutes of Health.

Hierarchical Data Format

Hierarchical Data Format (HDF) (hdf.ncsa.uiuc.edu) is developed and distributed by NCSA as a software library and format for the storage and exchange of scientific data. It is used worldwide in many fields, including environmental science, neutron scattering, nondestructive testing, and aerospace. HDF is also being used by NASA's Mission to Planet Earth and the Department of Energy's Accelerated Strategic Computing Initiative.

Symera

Symera (Symbiotic Extensible Resource Architecture) (symera. ncsa.uiuc.edu) has been developed by scientists at NCSA to take advantage of the proliferation of Microsoft Windows NT workstations in the research environment of most universities. It is a "distributed-object and cluster-management system with application support libraries built on Microsoft's Distributed Component Object Model (DCOM), designed for both parallel and serial applications" (Flanigan and Karim 1998). In short, NCSA has designed this software to allow combinations of common PCs to function together to gain computational power which approaches that found on more robust high-performance machines. As an example of this functionality, the development team has converted a stand-alone Windows program called Life3D to a Symera-style application. Similar work is being done at Northeast Parallel Architectures Center with their TANGO Interactive project.

Globus

Globus is a project of the Argonne National Laboratory (www.anl.gov), one of the U.S. government's oldest and largest science and engineering research laboratories. It received its commission as the first U.S. National Laboratory in 1946. It is operated by the University of Chicago for the Department of Energy with the general purpose of doing basic scientific research, managing environmental problems, and enhancing the country's energy resources. The laboratory has established the Distributed Supercomputing Laboratory (DSL) (www-fp.mcs.anl.gov/dsl) within its mathematics and computer science division. The DSL is developing the type of high-performance computing and networking model that the Alliance grid seeks to become. To assure that researchers and computational scientists gain access to this tremendous resource, it supports research and development efforts that result in high-level programming for parallel and distributed computing environments. Globus is one of its key programs (www.globus.org).

Globus is a prototype and a delineation of the fundamental technology that is needed to build the future computational grids. While Globus itself defines an application infrastructure to support the connection of computers on the Grid, GUSTO (Globus Ubiquitous Supercomputing Testbed) is the first generation network that is being used to experiment with, and fine-tune, the new protocols. When Ian Foster, Ph.D., of Argonne, and Carl Kesselman, Ph.D., of University of Southern California School of Engineering's Information Sciences Institute (USC ISI), received the prestigious Global Information Infrastructure (GII) Next Generation award for their work on GUSTO on April 20, 1998, the professors said

> Our hope is that access to distributed resources such as supercomputers will someday become consistent, dependable and pervasive. People will take it for granted. Imagine if the average small investor had access to a 10 million dollar supercomputer able to run a billion calculations per second. These are the tools that banks and the captains of industry have. Fundamentally new applications such as tele-immersion, remote visualization, smart instruments, and distributed supercomputing are only possible with the creation of new networking software (Globus 1998).

In a demonstration of the power that can be tapped with the Globus infrastructure, computer scientists at the California Institute of Technology used an application called SF-Express to link nine of the largest computers in the United States to run an interactive simulation that was a thousand times larger than any that had ever been run on a single supercomputer previously.

High Performance Virtual Machines

High Performance Virtual Machines (HPVM) (www-csag.cs.uiuc.edu/projects/hpvm.html) is an initiative by the Illinois High Performance Virtual Machines Project that seeks to increase "the accessibility and delivered performance of distributed computational resources for high-performance computing applications." They intend to do this with HPVM, a new software technology that takes advantage of the software tools and the accumulated experience many researchers have gained using traditional parallel computation models. The HPVM will take advantage of the increasing power of retail-level PCs running Windows NT or Linux operating systems to build scalable clusters, or groups of PCs, dedicated to a common problem. The effort goes beyond the commercial initiatives that have focused on multiple-processor, clustered workstations for the purpose of speeding up transaction processing on large data sets of commodities. It is intent on boosting scalable performance for a range of applications that include those which will form the basis for Internet2.

TANGO Interactive

TANGO Interactive (trurl.npac.syr.edu/tango) is a project that has been developed at the Northeast Parallel Architectures Center (NPAC) (www.npac.syr.edu) of Syracuse University, a research center that works on large information systems and simulations applications. TANGO Interactive is a Java-based collaborative system for use on the Web. NPAC takes the view that the Internet, and especially the Web, is the perfect infrastructure for cooperative interaction. TANGO has been shown to provide structure for Java-based applications in distance learning, remote consulting, support, and programmable 3-dimensional virtual environments.

Condor

Condor (www.cs.wisc.edu/condor) is being developed by a team at the Computer Sciences Department of the University of Wisconsin—Madison. Condor is a proposed solution to support high-throughput computing (HTC) on large collections of networked computers. This concept looks at the practical aspects of achieving large processing capacity, or a high level of floating point operations (intensive mathematical calculations), over an extended period of time. Floating point operations per second (FLOPS) has been the benchmark for high-performance computing across the research domain, but experimental scientists are now more interested in achieving HTC in support of higher quality computational results. Condor works to develop protocols that will allow diverse users on diverse systems to share their computational resources (regardless of the type of hardware or operating system) in a high throughput computing environment.

Virtual Environments

One method of making computing and collaboration more transparent and more "reality-based" is via virtual environments that will eventually allow for a total immersion of the user within data sets. In the near future, we should be seeing some very sophisticated machines that allow researchers to "reach-in" to their chosen environments (e.g., a human body) whether that is in the same room, down the block, or around the world. Currently, the Alliance is tracking the work of several virtual reality environments that have shown promise. Two, the CAVE and

ImmersaDesk, that have emerged from the Electronic Visualization Laboratory are profiled below.

The CAVE and the ImmersaDesk

The CAVE, which actually stands for CAVE Automatic Virtual Environment, is a virtual reality (VR) "room" developed by the Electronic Visualization Laboratory (EVL) (www.evl.uic.edu) of the University of

Figure 2.1. CAVE Immersive Environment.

Courtesy of Fakespace Systems Inc., Ontario, Canada. Source: www.pyramidvideo.com/products.html.

Figure 2.2. ImmersaDesk Virtual Model Display.

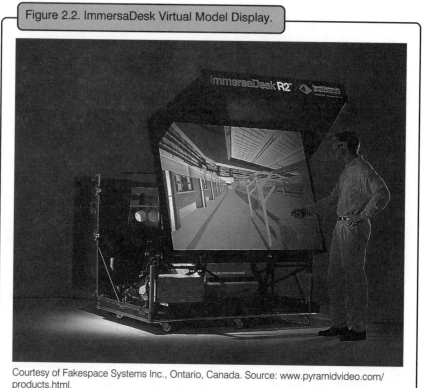

Courtesy of Fakespace Systems Inc., Ontario, Canada. Source: www.pyramidvideo.com/products.html.

Illinois at Chicago. A thorough description of the environment, including pictures and diagrams, can be found at www.vrco.com/CAVE_USER/index.html. The introduction to the user's guide found at that site, provides this succinct overview of the product.

> The CAVE is a projection-based VR system that surrounds the viewer with 4 screens. The screens are arranged in a cube made up of three rear-projection screens for walls and a down-projection screen for the floor; that is, a projector overhead points to a mirror, which reflects the images onto the floor. A viewer wears stereo shutter glasses and a six-degrees-of-freedom head-tracking device. As the viewer moves inside the CAVE, the correct stereoscopic perspective projections are calculated for each wall. A second sensor and buttons in a wand held by the viewer provide interaction with the virtual environment. The current implementation of the CAVE uses three walls and the floor. The projected images are controlled by an SGI Onyx with two Infinite Reality graphics pipelines, each split into two channels (VRCO 1998).

A second VR machine developed by EVL is being touted as the portable alternative VR environment. The ImmersaDesk looks like a drafting table, but operates as a virtual prototyping device with a computer operated audio system. Through the use of stereo glasses and magnetic head and hand tracking, the system offers a semi-immersive type of VR. Unlike the total surround effect achieved within the CAVE, the ImmersaDesk features a 4-by-5-foot rear-projected screen at a 45-degree angle. Second and third generation models can be packaged for shipment and quick set up for remote operation.

Software for the CAVE and the ImmersaDesk system have been developed to allow for networking remote users who want to collaborate on problems. In addition, a work station emulator is now also available.

Advanced Computational Infrastructure

The distinction of leading the research to develop the second advanced-computing prototype, the National Partnership for Advanced Computational Infrastructure (NPACI) (www.npaci.edu), goes to the San Diego Supercomputer Center (SDSC) at the University of California, San Diego. The Executive Summary of the proposal to NSF from the supercomputing center at San Diego states the goal of their challenging effort:

> The time has come to build a comprehensive, national computational infrastructure. High-performance computational technologies—supercomputers, high-speed communications, and storage systems—have been growing in capability and power at a rate unimaginable a decade ago when the NSF supercomputer centers program began. At the same time, digital scientific and engineering information is experiencing explosive growth. The evolution of computational capabilities and the growth in available data present the opportunity to revolutionize the national infrastructure. The National Partnership for Advanced Computational Infrastructure (NPACI) will lead the deployment and evolution of a national-scale metacomputing environment that will bring unprecedented power to bear on the challenges of computational science and engineering (NPACI 1998).

NPACI has similar goals to those of the Alliance, but the way the San Diego-led group is approaching the challenge is subtly different. The NPACI is composed of 37 different research institutions, located in 18 states across the continent, intent on implementing an information and collaboration network that supports a teraflops computer, a petabyte archive, and a metacomputing system that links important data servers at partner sites. The direction they have chosen to follow is contingent on the use of several scientific applications that are forming the basis of how the infrastructure will eventually be integrated.

"Thrust" Initiatives to Build the Computational Infrastructure

NPACI is beginning its work with a focus defined by two so-called thrust areas that reflect the people and expertise associated with the science community of their consortium. The endeavors of the applications thrust area, which includes engineering, molecular science, neuroscience, and earth systems science, will be supported by the work of the technology thrust. Included in this infrastructure-specific research are data-intensive computing, interaction environments, metasystems, and programming tools and environments. These areas of inquiry are explained below.

Data-Intensive Computing Thrust

Since digital data archives are evolving as an indispensable source of information for all scientific research, the development of a digital library system that can facilitate publication of scientific data sets, information retrieval, and data analysis services is important. This effort of the NPACI is led by Reagan Moore of SDSC but draws on the work being done at partner sites University of California (UC) Santa Barbara, University of California—Berkeley, Stanford University, University of Michigan, Washington University, University of California—Los Angeles, UCSD, the University of Maryland, University of California—Santa Cruz, Caltech and University of California—Davis. Massive data sets are only practical to analyze and manipulate if manual handling techniques can be eliminated. This is a goal of the data-intensive computing thrust area.

Interaction Environments Thrust

The interaction environments thrust is creating tools to ease scientists' access to information, analysis of data, and collaboration efforts across the network. Arthur Olson of Scripps Research Institute is the director of this project, which has the ultimate goal of providing easy access to the hardware and databases across the computational infrastructure. Once established, that access will allow for real-time data acquisition and collaboration through shared 3-dimensional environments. In the beginning, prototype efforts will attempt to integrate existing software tools into a visualization environment that optimizes scientific collaboration in the molecular science and earth system science thrust areas. As the results of these first efforts are evaluated, the group hopes to develop new tools to use in the neuroscience thrust. Brain mapping and the digital sky survey are two projects generating very large scientific databases, which can serve as a testing ground for the use of aural and haptic cues in communication and understanding of scientific data and models.

Metasystems Thrust

This thrust area is at the heart of the research on efficient ways to connect broad-based, broadband networks so that the unbelievably complex mix of hardware and software resources that reside on the different nodes of the network can be utilized as a single metacomputing environment. Metasystems, led by Andrew Grimshaw of the University of Virginia, is intent on developing the type of underlying infrastructure and easily used software that will make sharing resources across the distributed space a reality. Without transparent tools for access, the average user will find the network too complex to use. Metasystems is designed to provide these tools, which will result in the deployment of a virtual machine for accessing a single computational environment that shares processors, communication, and data. Initially, the metasystem will be tested by researchers within the applications thrust areas.

Programming Tools and Environments Thrust

The University of Maryland leads this thrust area. Headed by Joel Saltz, the research group intends to create tools to help scientists use the most powerful computing resources to solve scientific problems. The University of Texas, University of California—San Diego, the University of Virginia, Caltech, University of California—Los Angeles, University of California—Berkeley, Oregon State University, and the Center for Research on Parallel Computation will also help design testbeds to develop, evaluate, and disseminate experimental tools that will be used to share computing resources for problem solving. Objectives include

- Development of tools for irregular, adaptive, and unstructured problems.
- Creation of tools for data navigation and processing of multiresolution data sets stored in secondary and tertiary memory.
- Building of libraries to couple heterogeneous collections of parallel applications.
- Providing compiler support for out-of-core applications and adaptive problems.
- Development of parallel linear algebra libraries.
- Creation of infrastructure for solving linear algebra problems with library routines on computational resources across a heterogeneous network.

Demonstrating the Results

In April 1999, members of the Internet2 community got together in one of their regular scheduled meetings in Washington, D.C. Several projects operating under the auspices of the NPACI were invited to demonstrate their ongoing research and the results of their efforts to create the national information infrastructure of the future. One of these groups, Cooperative Association for Internet Data Analysis (CAIDA) Network Measurement, Analysis, and Visualization Tools (www.caida.org), showed several of their products designed to help network traffic measurement, analysis, and visualization. One of these tools is known as Otter.

The Otter application is helpful for visualizing arbitrary network data expressed as a set of nodes, links, or paths. An operator can use Otter to visualize Internet data that includes topology, workload, performance, and routing. The software can deal with any formatted data set that is composed of links and nodes.

A second presentation was done by the group known as MPIRE Interactive Viewing of the Visible Human (mpire.sdsc.edu). MPIRE, which is an acronym for Massively Parallel Interactive Rendering Environment, is a system designed to render multigigabyte volume data sets. MPIRE uses a graphical interface that allows the user to control the rendering process on a remote, massively parallel supercomputer via a Web browser. The project being demonstrated at Internet2 used large data sets from the National Institutes of Health (NIH) National Library of Medicine's Visible Human project. At realtime interactive rates, data is manipulated and delivered to a remote computer via the Web.

Virtual World Server/Virtual Los Angeles (mml.cs.ucla.edu) from UCLA's Architecture and Urban Design Department is a virtual reality environment, which is projected to grow as large as multiple terabytes and is accessed via a client-server system. In the demonstration for the meeting attendees, a client in Washington, D.C., was used to access the server in Los Angeles. Then, in real-time simultaneous rendering, the Virtual Los Angeles database was navigated.

References

EVL. 1998. "Art and Science Educational Programs." Electronic Visualization Laboratory (EVL), University of Illinois, Chicago. http://www.evl.uic.edu/EVL/RESEARCH/art_science.shtml. (8 June 1998).

Flanigan, P., and J. Karim. 1998. "Distributed Computing." *Dr. Dobb's Journal,* November 1998. San Mateo, CA: Dr. Dobb's Journal. http://www.ddj.com/ddj/1998/1998_11/LEAD/LEAD.htm. (12 December 1998).

Globus. 1998. "The Globus Project." http://www.globus.org/. (1 June 1998).

NCSA. 1998. The Alliance. ACCESS and the National Center for Supercomputing Applications (NCSA). http://www.ncsa.uiuc.edu/alliance/alliance/. (10 July 1998).

NCSA. 1999. "What Exactly is the Grid?" ACCESS and the National Center for Supercomputing Applications (NCSA). http://access.ncsa.uiuc.edu/CoverStories/WhatisGrid/. (8 February 1999).

NPACI. 1998. NSF Proposal, Executive Summary." National Partnership for Advanced Computational Infrastructure (NPACI). http://www.npaci.edu/About_NPACI/exec-sum.html. (3 April 1998).

NSF. 1995. Directorate for Computer and Information Science and Engineering." Naitonal Science Foundation (NSF). http://www.cise.nsf.gov/. (2 September 1998).

VRCO. 1998. "CAVE User Guide." VRCO, Inc. http://www.vrco.com/CAVE_USER/CAVEGuide.html. (30 September 1998).

CHAPTER THREE
Trends for the Consumer

W hat has changed for the consumer of information-based products in the last three years? It would be simpler to ask, what product does *not* now contain an information component? What has *not* been made faster, cheaper, and more reliable? And what is *not* capable of being connected to a network of other information-based, smart machines? Moore's Law predicted in 1996 that processor power would double *twice* in three years, and it has. That reality has driven the development of new hardware and new applications into ever-expanding areas. Given that reality and the economic fact that fortunes can be made in weeks at companies that are technology oriented or have a ".com" in their names, the pressure to introduce new products and services based on the latest innovations is unprecedented. Basic research, new scientific breakthroughs, and readily available investment capital set the stage for an explosion in digital products and services.

Only a few of the most important changes will be covered in this chapter, since a complete overview of the improvements to the vast array of consumer-based applications and appliances would occupy a second book. And, what is written on this subject will probably be of more historical significance than of cutting edge reporting, because computerization is moving so quickly. The analysis that follows will concern itself with the most important trends in information-based applications and products that have emerged between 1996 and 1999. Where trade-marked products or registered companies are mentioned, their inclusion is only for illustration. These examples, while important for understanding how research has been applied to the consumer market, in no way constitute endorsements or recommendations to buy.

INTERNETWORKING

Without question, the development that has had the most influence on the design of digital machines and their application is the rapid growth in connectivity options. Today, many machines have the capability of connecting to one type of network or another. These options include connecting to simple local area networks (LANs) that are based in businesses or in modern home environments, or connecting via a TCP/IP (transmission control protocol/Internet protocol) standard to the Internet and thus also to other devices on the Internet. These include an assortment of machines, some of which have never before been considered "networkable." In the past, the personal computer and attached peripherals like a printer, scanner, removable storage drives (CD-ROM, DVD, ZIP, optical storage medium, tape, etc.), and even a digital or video camera have been sold as "connectable" or "networkable." But today, add to this list of networkable devices the network-ready personal digital assistants (PDA) that provide portable access to calendars, contact files, and e-mail; personal communications devices (PCD) like cellular and satellite phones; car-mounted Global Positioning System tools (GPS) that keep navigators on the right road via satellite positioning data; television set-top boxes; and almost any home appliance one might imagine.

A somewhat bizarre example is the Screen Fridge, a collaboration between Frigidaire and a British software company known as ICL. This prototype appliance, introduced in early 1999, incorporates a PC onto a refrigerator door. This appliance cannot monitor the contents inside the box, but the system is equipped with a bar code scanner, which allows the owner to scan in the last bottle of beer or a can of soup so a shopping list can be generated for printing or e-mailing to the store. It is connected to the Internet for full-featured browsing and e-mail. The PC comes with a television tuner also, allowing the cook to keep up with daily television events as meals are prepared.

This, however, is probably not the path that will be followed by most manufacturers. Now that Sun Microsystems' Consumer and Embedded Division has released Jini, companies have a simple way to expand networking capabilities out of the office environment and into the home. Jini is a set of class libraries that work in Sun's Java programming language and run on various Java virtual machine applications, and it has the potential of becoming the standard by which virtually any appliance can be plugged into a network. Mike Clary, Jini project director explains, "We're interested in device-to-device, device-to-computer and device-to-e-network connectivity, Jini gives you true device-sharing. We haven't found anything else as comprehensive as what we're trying to do, from a networking standpoint" (Foley 1998).

This last point is an allusion to the efforts of other companies as they attempt to control the new de facto standard for information appliance connectivity. For example, Microsoft Research Group is attempting to implement Bill Gates' digital nervous system concept of ubiquitous connections. This effort, which could leverage the set-top box technology of Microsoft's subsidiary WebTV Networks Inc., is named Universal Plug and Play. Meanwhile, Hewlett-Packard Co., is working on JetSend technology, and Lucent Technologies Inc. is experimenting with Inferno, two products which are similar to Jini. But Jini is in the market today with code available to any Java programmer who wants to add a communication component to a phone, a printer, or even a toaster. By 2001, Jini developers expect to see their technology so widespread that connections will be made simply and seamlessly. Any device that is plugged in will automatically share the network dial tone and be ready to work in a remote condition. At the outset, major manufacturers announced they were ready to support Jini. Those companies included Hewlett-Packard Co., Canon Inc., Palm Computing Inc., Motorola Inc., and cellular phone makers Nokia and Ericsson.

New Standards Are Introduced

PC manufacturers have started taking advantage of new, higher speed data paths between their motherboards and the many peripherals that are attached to them. Two protocols that have been on the drawing board for years have finely gained widespread support. The Universal Serial Bus (USB) and IEEE 1394 (also known by the name it's developer, Apple Computer, gave it originally, Firewire).

The first new standard, USB, is a universal peripheral connection specification that was designed jointly by Compaq, Digital, IBM, Intel, Microsoft, NEC Corporation, and Northern Telecom to eliminate the need to install add-in cards and software drivers and to activate a true one-size-fits-all plug and socket connection. Most hardware vendors now support the standard, and adapters are available to connect to older devices that used the slower serial, parallel, and accessories desktop bus (ADB on the Macintosh) port options. The conversion to USB should be complete for new systems by 2001. USB connections are "hot swappable," meaning that the computer and peripheral can still be powered when a USB connector is attached or released. In addition, as many as 127 peripherals can be daisy-chained off one port, and all of them can be used at one time. Data transfer rates can reach up to 12 megabits bits per second (Mbps).

IEEE 1394 (Firewire) was invented by Apple Computer in the first half of the 1990s. It was adopted by the Institute of Electrical and Electronics Engineers (IEEE) as one of the industry connectivity standards in 1995. The standard can support data transfer rates up to 400 megabits per second. Because of its very high bandwidth potential, it is the preferred transfer method for video and professional level cameras. As more PCs emulate Apple's Power Mac G3 Tower by placing Firewire on the motherboard, more devices like hard drives, removable storage systems, and set-top boxes will be manufactured to take advantage of the faster throughput. While USB provides a more than adequate improvement in the speed of data flow for printers and most scanners, IEEE 1394 should become the high-end standard when real speed is necessary. Compaq Computer Corporation, Matsushita Electric Industrial Co., Ltd. (Panasonic), Royal Philips Electronics, Sony Corporation, and Toshiba Corporation joined with Apple in February 1999 to affirm their support of the high-speed IEEE 1394 digital interface. The consortium has also decided to form a patent pool to efficiently license patents required to implement the standard.

As Jerry Meerkatz, vice president of Compaq's Commercial Desktop Division said at the time of the announcement,

> Compaq supports the wide adoption of IEEE 1394 technologies and the efforts of the 1394 patent consortium. We believe that this alliance will ensure acceptance of this standard throughout the personal computer and consumer electronics industries by making the licensing of the key technologies easy and affordable. The world is evolving into a digital-based workplace and these technologies will allow for broader connections between digital consumer electronics equipment and personal computers, making it easier to use products (Apple 1999).

No Slow Connections

The advent of increasingly simple and speedy connectivity that initiatives like Jini, Universal Plug and Play, USB, and Firewire have produced speaks to the importance manufacturers have placed on implementing a distributed computing model. The trend is defined by new, simple, fast, and ubiquitous connections that rely on the TCP/IP protocol, the language of the Internet. As this new infrastructure becomes established, it is also the basis of enhanced multimedia capabilities that have signaled the acceleration of the convergence of computer, television, radio, and telephone media. The world is fast becoming digital. Voice data inter-mixed with video data intermixed with e-mail text can all be sent down the same carrier (wired or wireless) and eventually decoded at the end of a transmission that happens in seconds. Getting this capability to the average consumer at home is a technical challenge that has progressed in recent years.

In 1996, the standard by which a home-based PC user could expect to connect to the Internet was via a modem that supported peak throughput of 28,800 (28.8) baud, or approximately 28.8 kilobits (Kb) per second. This speed jump (from the earlier 9,600 and 14,400 (14.4) baud of the early 1990s) increased the bandwidth capability for those connecting to the Web, and made it a viable medium for larger graphics and streaming video and audio applications. Research, collegiate, and corporate users have enjoyed much higher bandwidths, of course, connecting not by the analog Plain Old Telephone Service (POTS), but via digital services (old standard T1, [digital carrier line capable of 1.544 megabit service] and high-bit rate DSL [digital subscriber line]) that move data at 1,544,000 bits per second. At these rates, real-time videoconferencing, for instance, became a very useful service. But it would take improvements in software code and enhanced modem compression techniques before applications such as these might be considered for home use.

The adoption of the V.90 international standard, which allowed for a download speed of up to 56 kbps (kilobits per second) and upload speed of 33.6 kbps, was an important agreement reached by the International Telecommunication Union in 1998. This ended a year-long fight for market dominance between 3COM and its US Robotics X2 56 kbps protocol and a competing standard that achieved 56 kbps speed, which was used by Lucent Technologies and others. By the end of 1999, the 56k modem was being used by approximately 100 million users in North and South America. The 56k-V.90 standard has improved speeds using the existing copper already on the telephone pole which means nearly every home in the United States could theoretically use these modems.

But the new standards on these latest analog devices are still not the ultimate performance enhancement for connecting to the Internet. In the first place, the Federal Communications Commission (FCC) has decreed that top speeds for data passed from the Internet Service Provider (ISP) to the modem should not surpass 53 kbps. Speeds on the return trip are closer to half that peak capability. And if a user is not blessed with superior line quality or is hampered with a very old phone line or connections, signals will be degraded markedly.

Because of these shortcomings, other options for broadband access have been tested and are now starting to be deployed commercially, albeit in limited areas. The most viable of the alternative connections comes from the cable television industry, which has spent millions of dollars throughout the United States to upgrade its infrastructure in order to send digital signals more efficiently. Most cable firms have replaced their old coaxial cable with fiber optic cable in the main trunk portions of their lines. This means that video signals can coincide with Internet traffic moving from the head end to subscribers' homes. By installing one of the new cable modems, the end user now has access to data flows that move in the range of 1 megabit to 38 megabits per second—that is, if the service is available in that particular area, and if the number of subscribers upstream is not too large. Cable Internet service is in the early stages of deployment in 1999, but it does show promise if the companies can establish high-quality service. Working toward this end, cable modem vendors Cisco Systems and @Home Networks have recently announced an agreement with Intel to develop modems that support both USB and Firewire connections to PCs, so that data bottlenecks will be alleviated and connection setup, theoretically, will be less painful.

A well-publicized takeover of the largest cable company in the world, Telecommunications, Inc. (TCI) by telephone giant AT&T in 1999, illustrates just how important the potential of cable is to the telecommunications (or telco) industry in terms of the deployment of digital services. Predictions by some experts say that cable will take an 80% share of the broadband market by 2002. AT&T expects that it can become a powerful onramp for consumer access to the broadband network through cable if its investments in other areas, like xDSL don't meet expectations.

xDSL is a general designation for a wide array of digital delivery options under the digital subscriber line (DSL) group of technologies. xDSL still uses modems to send data over existing copper wire. The original DSL was ISDN, which was developed in the early 1980s to obtain 160-kilobit-per-second rates. And HDSL (high-bit rate DSL) was the protocol that replaced what had been known as the T1 connection. The most common protocol being made available today is Asymmetric Digital Subscriber Line (ADSL), which has the potential to provide rates well above the standard 56-kilobits-per-second level. Unfortunately, while standards have now been set by American National Standards Institute (ANSI) to cover ASDL, it has not been deployed in as many geographical areas as was the promise in recent years. The issues behind the standards, deployment, and technologies of xDSL are intriguing but complicated. A good reference point for further reading on this subject is a white paper (a policy paper) provided by chip maker, Analog Devices, Inc. available at www.analog.com/publications/whitepapers/products/xDSL.html.

An interesting development in this area has to do with two competing protocols being optimized for business application, Fiber-based Gigabit Ethernet and Fibre Channel Arbitrated Loop (FC-AL or in this discussion, Fibre Channel). Both are high-speed broadband technologies that have been developed over the past couple of years. Gigabit Ethernet has just been adopted by the IEEE as a standard (802.3z) that enables data traffic to be sent over fiber-optic cable at 1 gigabit per second. Relying on the 25-year-old Ethernet, and the more recent Fast Ethernet, technology with its speeds of up to 100 megabits per second, the protocol allows for relatively simple installation and migration within existing business LANs.

Fibre Channel is an emerging network technology that is designed for storage and server clustering. Boasting a transfer speed of up to 100 megabits per second per channel (and up to 1 gigabit per second in later versions) at distances of up to 6 miles, Fibre Channel can handle a maximum of 126 devices attached via hot swapping per loop. While Gigabit Ethernet is strictly a network infrastructure that requires data to ride an Ethernet packet, Fibre Channel is designed to be more adaptable and independent of any preset protocol. In this way it can support audiovisual applications, online transaction processing, data warehousing and mining, plus Internet and intranet browsing and serving. It is also being touted as a replacement for SCSI (Small Computer System Interface), the interface that has connected computers to storage media and other peripherals at traditional maximum speeds for many years. Predictions are that half of external storage devices will be connected via this protocol by 2001. The standard is supported by Sun Microsystems, Inc., Digital Equipment Corp., Hewlett-Packard Co., Data General Corp., and EMC Corp.

How this jousting between carrier platforms finally settles out may ultimately be influenced by the current work of several standards organizations now wrestling with the technical aspects of utilizing Internet protocol (TCP/IP or IP for short) as the carrier protocol. Much of the work being carried on to move IP into new areas is the responsibility of the Internet Engineering Task Force. In the next year, progress should be made on standardizing IP over the new technologies of Fibre Channel and Firewire, as well as cable and xDSL.

Internet Environment

Next Generation Internet initiatives like the Supercomputing Alliance and the Abilene Projects, mentioned in the previous chapter, will serve as experimental test beds and roadmaps for twenty-first century deployment of computing environments that will change the pace of life on Earth. And a surprising statistic may help to prove that this change is happening today. Sometime in 1998 data traffic exceeded voice traffic for the first time. Probably the most important change that has taken place in the digital domain in the last three years is connected computing. An estimated 100 million users have embraced what former Intel CEO Andy Grove has dubbed "an environment." He, like other experts, once made the mistake of dismissing the Internet as just another application. It must be obvious to even the most technologically sheltered by now that the Internet changes everything.

Respected technology reporter Dan Gillmor has analyzed Grove's vision of the future:

> As we surge toward a new millennium, the Internet has become more than the overwhelming reality of the technology industry's current existence. It is the foundation for the Information Age, the environment in which we will all be living before long.
>
> In Silicon Valley and tech centers around the world, smart people are creating the machine tools of that new age: the building blocks of communications and commerce in tomorrow's inter-networked world. It's surprising enough to see how quickly the Net is subsuming all that has come before, but it's beyond stunning when we contemplate the impact.
>
> The computing industry is morphing into telecommunications, and vice versa (Gillmor 1998).

This convergence of the traditional media and communications systems into an interactive digital/telephone/television/Internet hybrid is taking place at such an accelerated rate that no one is willing to guess which new model will dominate in the long term. Well-publicized corporate mergers are telling evidence that the landscape is, indeed, going to change. AT&T is intent on bringing high-speed digital data to homes through their merger with Telecommunications Inc. (TCI). The alliance between Teleglobe Inc. and Excel Communications Inc. took place in order to extend Internet access and long-distance phone services to a global audience. Vodaphone, Britain's largest cellular phone company will takeover AirTouch of San Francisco to establish a worldwide wireless network.

In related moves, Cisco Systems, a main source for data packet router and switching hardware on networks acquired Selsius Systems and Summa Four Inc., two companies that make voice equipment. Cisco plans to be a major provider of equipment that will change voice transmission into data packets. Lucent Technologies, looking to also be a player in data delivery, bought data networking companies including Prominet Corp.

and Yurie Systems Inc. Not to be left out, Microsoft Corp. has announced that it will invest $500 million in Britain's NTL Communications Inc. to develop a high-tech communications network in the United Kingdom and Ireland.

The reason behind this merger mania is economics. The development of technology and infrastructure that will eventually replace the telephone networks, broadcast television, and the Internet as we know it is extremely costly. As big as it is, AT&T, for example, does not have the resources to do the research, lay the lines or launch the satellites, and design the protocols that will be required to build the next generation broadband network. But in conjunction with TCI, they make a giant step toward completion of what experts call "the last mile": a connection to residences and home offices.

TYPE IT OR SAY IT

As the convergence of diverse media continues its migration to the Internet, one of the more interesting areas of development is "voice over IP." The technology that would allow one computer user to actually send sound from one machine to someone connected to a network across the hemisphere or even just down the stairs had its early beginnings in 1995. Then, without full-duplex capabilities, each user would be required to wait for the other to finish speaking (and transmitting) before they could begin speaking into (and transmitting from) the microphone that was connected to their computer. The quality left a great deal to be desired with delays, dropped words, and a lot of noise being the nature of the connection. The big advantage, of course, was that it was a free long-distance call.

Since then, voice transmission over Ethernet local area networks (LANs) and internal business calls over IP have become an alternative method of communication for many corporations. In the latter case, voice connections are most often done within the secure environment of a Virtual Private Network (VPN). These private and encrypted connections developed in 1998 as a service of Internet Service Providers (ISP) who saw a need to replace the old remote access protocols that had used direct dial-in capabilities in the past. VPNs transfer information by encrypting and encapsulating data in IP packets that are then sent over the Internet. This is called tunneling. On one side of this tunnel is a firewall that encrypts the data to be sent. On the receiving end of the tunnel, another firewall receives the packet, decrypts the message, and passes it to the receiver's PC.

The message can be in any binary format. In most cases, e-mail is still the overwhelming traffic being carried, but advances in voice decoding have made conversations a real option. This said, no one is claiming that the quality of voice transmissions that are carried over IP yet rival that of the established phone services, including cellular. A new product from Symbol Technologies was recently introduced to take advantage of the inroads in the convergence of media that have recently taken place. That company's NetVision Data Phone combines voice communication, data capture capabilities, bar code scanning, embedded Thin Client, and a radio card into a single handheld device about the size of a cellular phone.

The unique device is able to convert analog voice signals into compressed digital packets that are sent via TCP/IP over the data networks. At the same time, it can still use an existing phone system using standard features like call waiting, call transfer, and hold. The user can talk, hit a button, and send data, all during the same connection. The Thin Client includes the capability to browse the Internet.

LEAVING THE WIRES BEHIND

The efficiency of the NetVision phone device is based in large part on another trend affecting the distribution of data around the world. It relies on wireless technology that has come a long way from the days when radio was the only talking box in America's living rooms. Advances in the technology for sending data via high-frequency radio waves, allowing users to connect to a LAN, a fax, e-mail, or the Internet are being made on several fronts:

- Cellular Digital Packet Data (CDPD) is an IP-based network that sends and receives data over the cellular spectrum with a transfer rate of 9 kilobits per second. To date, 55% of the United States is able to access the service.
- Packet radio systems are the protocol in use by a national network built by Ardis and BellSouth Mobile Data. With transfer rates of up to 19.2 kilobits per seconds, the system is mostly used for data messaging and other specialized applications.
- Personal communications services (PCS) were originally built for digital paging, but were modified to include voice capabilities. Different protocols are currently being deployed under the PCS umbrella. Global System for Mobile Communications (GSM) is a network for digital cellular that supports voice and data connectivity. It uses a technology that is able to divide radio frequencies, so that it can carry multiple data channels. GSM is the European

standard. Code Division Multiple Access (CDMA) is the cellular standard and the most often deployed of the technologies. It allows for the reception of digital data, but the user cannot send data.

Only about 5% of people connecting to the Internet currently do so by some form of wireless modality, but that figure is expected to rise rapidly. As coverage across the country becomes more universal and if competing companies can stabilize on a competitive, cost-effective pricing structure, the protocols will certainly be ready to carry the weight. Evidence of this trend is seen in some recent developments across the industry. For instance Lotus has released a wireless version of Lotus Notes that includes an intelligent agent feature for automatic updating of mobile databases and for filtering e-mail. Another company called Unwired Planet has joined with wireless phone manufacturers Ericsson, Nokia, and Motorola to form the Wireless Access Protocol (WAP) Forum. The objective of the group is to develop a standard for writing applets that optimize IP over wireless networks. Qualcomm is working on an implementation of the third generation cellular network standard, which may eventually see speeds that rival T1 standards in wired lines.

Microsoft also leads an initiative in the wireless domain with Sharp and Hewlett Packard, establishing the Windows CE group, with the goal of expanding wireless implementations of Windows CE applications. Microsoft has also gone ahead with a joint development deal with Qualcomm, called Wireless Knowledge. The services established by this joint venture are expected to include access to e-mail, calendars, contact lists, and connection to Microsoft Exchange servers. In an approach that the company terms "end-to-end Windows data access," the access will be wireless protocol-independent. Connections will be supported over CDMA and GSM, and links can occur between a wide range of devices, including Web browser kiosks, Web-enabled televisions, and Windows laptops.

Concurrent with all this activity, Bell Labs once more announces a breakthrough in another one of its data flow research initiatives. Their demonstration of a "no-fiber optical data link" may prove to be a far more efficient method for transporting bits than is available in currently deployed commercial wireless data links. The present standard connection speed is 622 megabits per second (that is 622 million bits). Their new procedure set a world record by transmitting data 1.5 miles through the air at the rate of 2.5 gigabits per second (that is 2.5 billion bits), error free. The technology, which incorporates two custom-built telescopes with optical transmitters and receivers and a high-power optical amplifier, may prove to be one of the solutions for high bandwidth deployment in locations where cable, telephony, or other traditional communications infrastructures are not available and/or are too costly to build.

THE PROGRAMMING LANGUAGE OF THE INTERNET

Java was first developed in 1995 with pronouncements that it would likely replace Windows as the dominant operating system when all computing moved to the distributed architecture of the Internet. Nearly five years later, Sun Microsystems' Java language has yet to fulfill this lofty prediction, but progress toward a "write once, play anywhere" computing environment has been made. In December of 1998, Sun released the latest incarnation of its software development package, Java 2. The new Java addressed some security and interoperability questions that arose with the earlier version. Boasting faster performance, Java applications should now work better on all of the computing platforms that support the Java Virtual Machine environment. Much of the speed enhancement, according to analysts, is by virtue of Java accessing the host computer's operating system to use features like graphics rendering and drag and drop transferring of data.

Almost a million programmers have licensed Java code, so they can build and distribute computer applications in Java. The latest version has been opened up to include a development environment for noncomputer applications as well. This means that Java will likely run some applications and functions of common machines like the automobile. And as was mentioned previously, the Jini class of libraries in Java established a de facto standard for operating and connectivity software for the emerging information appliance market. Sun has even announced plans to develop a Java optimized chip to run the software that will be created for devices like television set-top boxes and the like.

The other important advance that concerns Java is the altering of the developer licensing model, which has now been combined with the open source standard. While Sun continues to hold the patent on its language, it has freely distributed the source code so that independent developers have the option of altering the basic framework to make it run in a more optimal manner, depending on the function involved.

As the challenges of alternative operating systems, such as Java and, later, Linux became apparent, Microsoft developed Windows NT, an operating system that is distributable across a network (be it a LAN or the Internet). Windows NT is constantly being improved, but critics claim that it is too large and unstable to be the operating system of choice. Windows CE, which was developed as a mobile computing option to run applications on personal digital assistants (PDAs) and other small appliances, could hold more promise in this area. Sun may well have positioned itself as the environment of the future with Java, Jini, JAIN (Java Advanced Intelligent Networks) and JavaSpaces. However, no one should count out Microsoft.

MOBILE COMPUTING

As corporate America changed its organizational models to reflect a more decentralized, open design that allowed workers to untether from their traditional office/cubicle/desk environments, the mobile computing options exploded. This trend, while obvious as early as 1995, is still emerging. A 1995 study estimated that only 265,000 workers were using devices that would facilitate working and accessing information remotely via portable devices, which might include a notebook computer, a subnotebook, a pen-based system, or a personal digital assistant (PDA). In 1998, the PDA sector of the mobile computer industry alone accounted for sales of 3,900,000 units.

Much of this growth is the result of the popularity of 3Com Corp.'s Palm III (and it's predecessor, the Palm Pilot), a system for mobile workers who want to stay organized (calendar/contact/note-recording databases) and in touch (via e-mail, fax, Web), using a device that they can carry in their pockets. Almost 41% of handheld devices shipped in 1998 ran on the Palm operating system, 25% used the Windows CE system, and 13% ran Psion's operating system. Each of these options boasts connectivity as well as productivity applications that are often compatible with full-blown desktop versions. And for those who use handheld units like the Palm III, the Phillips Nino, or dozens of other devices, or a smart phone, which can combine some PDA functionality with cellular voice capabilities like the Nokia 9000il, many also find it useful to carry a notebook computer.

Over the past three years, the typical notebook (also known as a laptop) computer has dropped in price from a range of $3,000 to $7,000 to something in the neighborhood $1,500 to $2,000 for processing and storage capabilities that have at least doubled. As has been stated before, Moore's Law predicts this magnitude of improvement. The standard for a unit priced around $2000, as of the spring of 2000, was a Pentium III-based processor running at 500–600 megahertz (MHz), 64 megabytes (MB) of RAM, a hard drive of 6 gigabytes (GB), a 56-kilobits-per-second (kbps) modem, and a CD-ROM/floppy drive option.

NEW PRODUCTS ON THE HORIZON

What will be available in the near future can only be guessed, but highlighting features that may prove to be useful adaptations of the current technology may shed light on the probable possibilities. Three devices released in late 1999 or early 2000 are good examples of the direction one can expect to see exploited. The first is the latest in the Palm family of handhelds, the 3Com Palm VII.

The latest incarnation of this popular personal digital assistant (PDA) will now feature two-way wireless connectivity for both messaging and Web access. The price is expected to be double that of the current model, which retails for about $400, and it will require a subscription to a proprietary Internet service (3Com's own Palm.Net), which will run about $10 per month. After the owner deploys the flip-up antenna, a micro-sized radio transmitter will be able to connect to the Bell South wireless data network almost anywhere within the borders of the United States. Plans are in the works to broaden the network to include international coverage. To lower the amount of data transmitted and received, Palm.net is partnering with dozens of content providers who will be modifying their original Web pages to include the most pertinent data, optimized for viewing on the Palm's smaller color screen. To date, the content providers include ABCNews.com, Bank of America, ESPN.com, E*Trade, Fodor's, MapQuest, MasterCard, Merriam-Webster, Moviefone, TheStreet.com, Ticketmaster, Travelocity, UPS, USA Today, US West, Visa, The Wall Street Journal Interactive Edition, The Weather Channel, and Yahoo!

Chip maker Cyrix is also offering an interesting machine. The product is WebPad, a 3-pound, 8.5-by-11-by-1.75-inch, battery-powered tablet that contains a Web browser and a 12-inch color touch screen. It will be offered with an included stylus (optional keyboard and mouse are available) and an onboard transceiver. The unit will connect to a base station via radio frequency, much as a portable phone works today. The base station is connected to an analog modem, ISDN, xDSL, or Ethernet line from which the Internet can be accessed.

But if the impressive capabilities are still too limited for the person who just has to have the ultimate connection, the wearable PC may be just what the doctor ordered. Earlier attempts to make a personal computer powerful and small enough to carry on the human form have met with limited success. Either the power wasn't there (both in processing and in battery life), the devices were too cumbersome and bulky, or the output solution was not adequate. MIT's Media Lab has been working on these problems.

Their current version of a wearable PC is about the size of a small loaf of bread which is worn on the waist or over the shoulder on a strap. The CPU is accessed by a Twiddler, a one-handed keyboard that can be manipulated via a series of small buttons that have to be learned. A head-mounted monochrome display that covers one eye is used for readout. The machines are being tested by graduate students who travel to, and attend their classes with the wearables attached. They take notes, perform common computer tasks, and generally act as normal students to see how the machines can be improved. One idea they are now considering is the creation of software-based autonomous agents or "long-lived applications that reside in the background of your computing environment and perform a given task without supervision." An agent might be programmed to keep track of Web pages being accessed so that it could make suggestions of related sites or point to already archived files. And since the wearable's ultimate utility will be demonstrated by the way it helps the users interact with their immediate environment, agents could be trained to take in conditions that might be affecting the users so as to make suggestions about behavioral changes that could optimize their experience.

Xybernaut Corp. and VIA Inc. are two companies pioneering the commercial application of the wearable PC. They have recently introduced versions that are almost comfortable. Early adopters like engineers at Ford Motor Corp. and runners on the New York Stock Exchange are experimenting with the VIA system. And, we can expect to see camera operators covering the Super Bowl outfitted with the Xybernaut gear. By 2000, one of the largest of computer corporations will demonstrate that this trend is not just a passing fad. IBM will deliver its Visionpad based on the company's ThinkPad 560 notebook. The system uses a 233MHz Pentium MMX and includes 64MB RAM and one of IBM's own microdrives, a 340MB flash memory card. A head-mounted display will provide 800-by-600 resolution and 16-bit color, and input will be achieved mainly through ViaVoice voice-dictation software (a USB keyboard will be supported, but it is extra). Total weight for this package is 10.5 ounces.

MASS STORAGE ADVANCES

Since the first computing devices were created, the storage and access of data has been accomplished in generally the same manner. The platters of aluminum alloy that make up the "hard disks" of the storage drive are coated with a magnetic material a few millionths of an inch in thickness. Data is written to and read from these surfaces, which are spinning at a minimum of 3,600 rpm via the read-write heads. These heads are located on an arm that moves over the platter at distances considerably smaller than the width of a human hair. When data is being stored, the computer software sends an electronic charge to the head so that a magnetic pattern can be laid on to the surface. In the retrieval mode, the read head will pick up the magnetic pattern and convert it to an electronic signal, which then gets amplified and decoded back to the original information that was stored. The code is based on the binary system of 1s and 0s which are stored as opposites in the magnetic polarity (positive/negative) of the disk drive surfaces.

Modern drives can contain more than one platter, and they can hold data on either side. Even though the drives record data in concentric circles that are numbered from the outermost edge of the disk in, they also feature the critical ability of holding that information in random order. The computer keeps a record of the address of where data is stored on a track, and the read head can slide to that exact spot in a millisecond. Again, in correlation with predictions based on Moore's Law, improvements in storage capabilities and capacities follow the "doubling every eighteen months" pattern. In 1995, it was common to see PCs on the market advertised with 4 to 8 megabytes of RAM and containing a hard disk drive of between 350 to 500 megabytes. Today state-of-the-art storage is remarkably inexpensive, and it is not uncommon to see a system advertised with 5- and 10-gigabyte hard disks. One survey of storage prices noted that a megabyte cost nearly $1.00 in the early years of the 1990s. In late 1997 the price had dropped to about 11 cents per megabyte. Just before the year 2000, the average is close to three cents.

This price drop and increased reliability have been the result of improvements in the technologies that are responsible for hard disk configuration. The areas that have affected how well these drives function include the following:

- Improvements have been made in the materials used to manufacture the drives. For instance, the latest devices now use superior magnetic films and magnetoresistive (MR) heads.
- Improvements in the mechanical capabilities have increased the disk rotation speeds, shortening data seek times.
- Newer semiconductors, like the digital signal processors, have increased speed of processing requests and data retrieval.
- Read-write signal processing improvements, like Partial Response Maximum Likelihood (PRMP), have made read channels more efficient.
- Improvements to the disk drive controller firmware and custom application-specific integrated circuit (ASIC) hardware have al-

lowed implementation of the faster control functions in data caching and error correction.

• Advances in the bus and disk interface have also been made.

Greater Storage Space Demand

Regardless of the advances in hard drive technologies, which have brought the price of 9-gigabyte drives to under $300.00 on the retail market, over 50 manufacturers are developing and shipping an array of removable data storage solutions that are based on different recording standards and interfaces. Each year, more imaginative (and mostly incompatible with previous modalities), higher capacity, and often smaller products are introduced. While a lot of the development has been driven by the need for better storage for new digital cameras and telecommunication systems, some of the removable devices are improvements to the older archiving functions that were once the domain of the floppy drive. The need for these magnetic rigid disk cartridge drives, PC Card rigid disk drives, floppy disk drives, small optical disk drives, and flash memory cards technology is growing. It was estimated by DISK/TREND, Inc. (www.disktrend.com), in its annual report on the state of the computer disk drive industry, that the almost $4-billion worth of all removable products shipped in 1998 will increase to more than $5-billion worth shipped in 2001. The following sections will present a brief overview of some of the improvements to the new removable storage options.

Flash Memory

These very small solid state memory modules have been commonplace in peripherals, notebooks, handhelds, and digital cameras since 1995, when the technology was needed as an alternative to the bulkier hard-drive and optical-drive solutions. The point of being mobile, of course, was to compact the size of the devices that needed to be carried. The array of different types of flash memory cards that now proliferate are the answer to this challenge. Research on ways to maximize memory capacity is the paramount concern of companies that market flash memory chips. A look at Intel Corporation, the world's largest chip maker, will demonstrate the advancements that have been attained over the years in this product class.

Their StrataFlash memory card recently has been improved to become a product that can hold two bits of data per cell instead of the usual standard single-bit-per-cell configurations. This higher density technology has given the company the advantage of being able to market a less expensive card in terms of cost-per-bit. As Intel co-founder Gordon Moore commented, "Two bits in the space of one starts a new direction in memory technology. This will lead to lower cost and new applications" (Intel 1997). The progress demonstrated by this new product's technological advance ironically undercuts Moore's own venerable axiom for predicting chip capabilities. While the law held from 1988, when the first 256-kilobit chip was introduced by Intel, to 1997, when the company announced the 32-megabit model, this new 64-megabit chip is double the prediction of where the technology should be.

Removable Optical Storage Options

The common format for removable optical storage systems is the compact disc (CD). In the early 1990s, the CD was a read-only medium, and today the CD-ROM (compact disc-read only memory) is still the ubiquitous choice for dissemination of large quantities of data. Movies, graphics, software programs, music "albums," and computer games are distributed by means of the CD-ROM. The drives that play these discs have improved considerably since 1995 when the standard was 2x (or two times the access speed of a common music CD, which is 150 kilobytes per second.) Today, it is most common to buy 12x drives as these have proven to be the most stable. However, 20x and even the new 32x drives recently introduced by Samsung Electronics are becoming readily available. The compact disc is able to hold approximately 640 megabytes of data on only one side of its platter.

In the past three years, improvements in the hardware that plays and records compact discs have lead to a great increase in the efficacy and marketability of two new storage formats, compact disk-recordable (CD-R) and compact disk-rewritable (CD-RW). The former is the more common today, a storage system (drive and media) that allows the user to transfer data from one source to the CD-R in what is known as a write-once format. That is, the CD is used for writing once. Data can not be altered or overwritten once it is "burned" the initial time, although a sector that was not used can still receive files. Reading of the information can occur many times with an expectation of zero data loss over a relatively long period.

On the other hand, CD-RW allows for the reuse of the media up to 1,000 times. Of the two, CD-R is less expensive and in more common use among home PC users. A shortcoming of these formats is that they are not interchangeable when being read. While CD-R is read-compatible with CD-ROM drives, CD-RW at this point can only be accessed in what are known as multiread CD-ROM drives. This problem is not a large stumbling block for most users who might want to back up their hard drives or make music CDs to listen to in their cars. And while the CD format will remain a standard for years, a new more advanced solution was recently introduced.

The digital versatile disc (DVD), which was originally known as the digital video disc because of its development for full-length movie distribution, is the heir apparent to versatile compact disc in the optical storage media sector. This platter is very similar in size and appearance to the CD, but it can hold several times more data than its predecessor. Since it can record information on both sides, depending on the layering technique being used, a range of space that occupies from 2.6 gigabytes to 17 gigabytes is possible. Like the compact disc, a digital versatile disc is available in different formats. DVD-ROM discs are the read-only version of the media. DVD-R discs can record only one time per disc. DVD-RAM and DVD+RW discs are two rewritable options that have evolved from competing branches of the industry. DVD-RAM, the initial rewrite format from Hitachi America Ltd., Toshiba America Information Systems, Inc., and Matsushita Electric Corp. of America, can hold 2.6 gigabytes of information per side. The other format, DVD+RW, can hold 3 gigabytes per side. Vendors supporting the DVD+RW standard include Hewlett-Packard Co., Philips Electronics, and Sony Electronics, Inc. One of the advantages of DVD drives is that they can read audio CDs, CD-ROMs, CD-R discs, CD-RW discs, and DVD-ROM discs. However, like the problem with the CD-RW, a disc created by a DVD-RAM drive won't work in a DVD+RW drive and vice versa.

Just as beta versus VHS kept many consumers on the sideline as new video players hit the marketplace in the 1970s, most users have been slow to adopt either of the rewritable DVD options. And the cost of CD-RW drives continues to be very attractive. At about $400, they are only double the cost of the less popular DVD-RAM drives. This price discrepancy is expected to last for several years, or until such time as the competing standards for the DVD writables are merged into one standard.

Where Is the Floppy's Successor?

Even into the year 2000, PC manufacturers have made the decision to include the 1.44 MB floppy drive as a standard feature on their machines. Given the reality of today's very large software programs, the common usage of high resolution graphics, video, and audio files, and the increasing size of the standard hard drive, that practice is likely not to continue. The floppy just does not have the needed capacity anymore. The only manufacturer who has taken the risk of moving beyond the legacy architecture is Apple Computer. When they introduced the ground breaking iMac in August of 1998, it not only eliminated the Apple desktop bus (ADB) and serial ports that had been standard on every one of their machines since the early 1980s, the floppy drive was also missing. Critics, while praising the operation and price point for this all-in-one internet-ready computer, wondered aloud about how consumers would react to this change.

By some accounts, there are still many users who need to load programs and retrieve or share data that only come on the floppy format. These legacy questions, it would seem, are the reason other companies have been reticent about cutting or changing out the floppy for a more modern, higher density storage solution. Not wanting to frighten customers, they have continued including hardware of dubious value. It is not as if superior solutions don't exist. Over the past several years, alternative removable storage devices have been offered by many manufacturers. Soon one of them will be chosen as the new standard.

Of the five competing (and mutually incompatible) products that are likely to receive the nod in the market place, three feature backward compatibility with the old standard. They can read 1.44 megabytes floppies. The likely list of floppy drive replacements include the following:

- The Imation SuperDisk, also known as the LS-120, has a capacity of 120 megabytes. Compaq, Gateway, and Hewlett-Packard currently offer a SuperDisk option with some of their desktop and portable computers. The drives cost about $100.00.
- Caleb Technology offers the UHD 144 with a capacity of 144 megabytes. It is available on desktop model computers only. Since it is late to this market, there is some question about its chances of being accepted by the major manufacturers. (Another company with same "late to market" hurdles is Swan Instruments with its UHC 3130 drive.)
- Sony HiFD is the entry from the company whose 3.5-inch floppy drive became the standard years ago. With a 200-megabyte capacity, faster data transfer rates, 3600-rpm rotational speed, separate heads for 1.44-megabyte and 200-megabyte media, this option may be the best performer on the market. Cost is about $200.00.
- Iomega has proven to have the most popular alternatives to floppy drive-based storage over the last four years. Its Jaz, and especially, the Zip drives feature a lot of room for a moderate price. The Jaz comes in a 1-gigabyte or 2-gigabyte version, and the Zip uses a 100-megabyte medium. Over 14 million of the latter drives have been purchased for PCs, Macs, or notebooks. They are, however, incompatible with the 3.5-inch floppy. If this

becomes the standard, the floppy drive will continue to be necessary in PC configurations for some time.

FACE TO FACE WITH THE INTERFACE

In the first chapter, the Artificial Intelligence section was devoted to some of the leading edge scientific laboratory research that is being carried on to make the computer a more "human" machine. Over the years, results from this and other computer-related inquiry have been quickly adapted by business interests in an attempt to make products that will appeal to the consumer. For instance, when Steve Jobs was starting Apple Computer in the early 1980s, he saw the results of work at Xerox's Palo Alto Research Center (PARC) that produced the first mouse and the first Graphic User Interface. Both of these features appeared in Apple Corporation's first Macintosh in 1984.

This connection between the research and the eventual application is still very much alive as the following descriptions of human-machine interface improvements will attest. Speech recognition in particular has made great strides in usability in the last few years. Digital dictation systems, especially, have gone from a novelty application to something that can prove to be practical. The big three companies in this area are Dragon Systems, Lernout & Hauspie, and IBM Corp.

Speaking to the Machine

L&H Voice Xpress is the name of the software package being marketed by Lernout & Hauspie, a pioneer in the voice recognition arena. The latest version supports dictation into virtually any Windows application. But its real power is displayed within Microsoft Excel, Word, PowerPoint, Outlook 98, and Internet Explorer where the interface provides a natural language command and control of these applications. The software, once the user goes through a training process, has the ability to operate the Microsoft Office Suite of programs with minimal contact with the mouse or keyboard. One can switch between programs by simply saying, "open Word," or "switch to Excel." And once inside Excel, for example, the application will respond to commands like, "sum the columns A, B, and D."

L&H Voice Xpress supports manual training, and a user can add specific words to the 30,000-word default vocabulary. A special feature allows the user to download recordings made with a remote dictation machine. This remote functionality is taken one step farther by the other giant of voice recognition, Dragon Systems.

Dragon NaturallySpeaking Mobile includes a hand-held recorder, a voice file transfer program, and the Dragon NaturallySpeaking award-winning speech recognition software application. Designed for people on the go, the package features a recording device which measures about 5 inches by 2 inches and weighs only 3.8 ounces. The recorder comes with a memory capacity of 40 minutes, but it is expandable with various flash media products.

After a 20-minute session in which the user records a training text to set up the program, the "recognizer" algorithm will spend another hour to adapt to the user's recorded speech. As the program operates, it continuously adds the specific vocabulary of the user and learns speech patterns. Accuracy, as a result, is improved. *PC Magazine* tested both the Dragon and the Lernout & Hauspie products to determine which was superior. Both scored exactly 11 errors out of a 191 word test.

The IBM ViaVoice software has also received good reviews of accuracy by users. As a computer maker, IBM has started to incorporate this software into its latest hardware offerings, touting the use of voice navigation tools to access the desktop and to use programs like their Lotus Notes. Like the other software mentioned above, ViaVoice comes with an integrated headset that includes a microphone.

In the case of ViaVoice, a speaker is allowed to dictate sentences at a rate up to 140 words per minute. This is slower than most people naturally speak, but this rate is far in advance of the discrete word systems of about two years previous. In that case, the user was required to pause a second between words making the dictation process a cumbersome and unnatural one. As the intelligent systems software of each of these companies continues to evolve and the processors get faster, voice recognition will be a feature in many information devices. Microsoft is actively pursuing integration with the next version of the Windows OS, and other designers are working on embedded chips that will eventually allow voice access/authentication of in-home devices that might activate security measures, start the VCR, or turn up the heat through voice commands.

Monitor and Screen Resolution Advances

Interesting developments are taking place in the visual interchange of data between the machine and its users. Of most importance to the average consumer must be the startling improvement in display technologies and the resultant price drop in high-resolution, large-screen monitors. Moore's Law-driven technological advance has been the hallmark of chip, memory, storage, and application development within the computer industry for many years. But when it comes to changes in the way

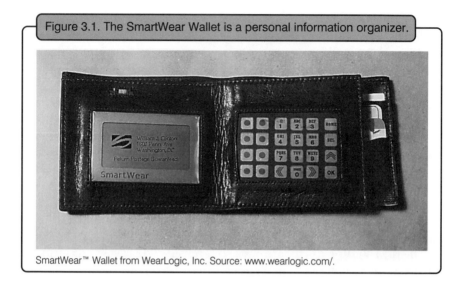

Figure 3.1. The SmartWear Wallet is a personal information organizer.

SmartWear™ Wallet from WearLogic, Inc. Source: www.wearlogic.com/.

users see information displayed, the improvements have not kept apace. That is, not until now.

In 1996, the only place one would expect to see a Liquid Crystal Display (LCD) monitor would be as an integral part of the notebook computer or on one of the many smaller devices like digital watches, calculators, and handheld personal digital assistants (PDAs). Since then, researchers have worked hard and long finding ways to improve the basic configuration of the LCD monitor so that prices could be reduced and its deployment could be expanded. The results of that work are now available in the form of new flat panel displays (FPD) that are thin, vibrant color, low-energy-consuming alternatives to the common cathode ray tube (CRT) monitor on most desktops today.

Unlike CRTs, the LCDs don't emit their own light. Some, like the digital watch, use ambient light that is reflected back to the viewer after it is polarized at the back of the display device. Other displays are backlit, using cold-cathode-fluorescent technology as a light source, or edgelit, wherein either electroluminescent or light-emitting-diode (LED) technology prevails. These displays can also be either transmissive or transflective in nature. Most notebooks use the transmissive mode, which features no reflected light in its deployment. It relies only on edgelighting or backlighting. The transflective option can use reflected light when it is adequate or backlighting as needed. Most PDA developers use the transflective option.

As manufacturers have improved these and other performance factors, some of the drawbacks originally associated with the LCDs have been ameliorated. A common complaint, for instance, was that it was hard to see the screen if one wasn't directly in front of the display. Many can now be viewed at more obtuse angles because of the practice of "in-plane switching." This function arranges the LCD horizontally instead of vertically. Manufacturers are also able to make thinner LCD layers that will react more rapidly to current changes. This allows for faster response times for playback of animated graphics and the motion associated with movies. And IBM is set to introduce an LCD with a resolution of 200 pixels per inch. The resolution is so good, say some, that it rivals the appearance of the printed page.

One of the most dramatic moments in Bill Gates' fall 1998 COMDEX keynote speech centered on the importance of creating displays that could mimic the quality of the printed page. In his introduction of ClearType, a software-based technology that is said to improve the readability of text displayed on an LCD monitor by 300%, it appears that he has done just that. Leveraging the work initiated by Apple cofounder Steve Wozniak in 1976, when he received a patent for a high-resolution display on the Apple II, the Microsoft ClearType breakthrough relies on a technique known as pixel splitting or subpixel addressing. Since a display's pixels are composed of subelements (green or purple in the case of the Apple II, and red, green, or blue in an LCD), manipulation at this level can generate smoother curves than the traditional anti-aliasing technique can accomplish. Steve Gibson, now of Gibson Research, was a programmer who helped develop the graphics and hardware for Apple Computer in the 1970s. He agrees that the process of subpixel font rendering is extremely effective in presenting text that is highly readable on an LCD monitor. Microsoft has plans to incorporate ClearType technology into its products by the turn of the century. If the reader would like to read an explanation of how subpixel rendering works today, Steve Gibson's site (grc.com/ctwhat.htm) is highly recommended.

One company that is not waiting for the new technology to arrive is NuvoMedia. They have calculated that there are probably quite a few consumers already sufficiently comfortable with the present LCD screen to read an entire book on a handheld platform. Their Rocket eBook is a 22-ounce device that appears to be a common personal digital assistant, (PDA) but is really a machine specifically designed to store the equivalent of 4,000 pages of words and graphics. By placing the eBook into its specially designed docking station, the user can connect to the Internet site of NuvoMedia or an online book retailer like Barnes and Noble and download an electronic version of several of the hundreds of books now available. The e-versions of the manuscripts are comparable in cost to the printed ones, and the device itself must be purchased for around $300. As of yet, the Rocket eBook is not a serious threat to traditional paper-based publications, but advances like ClearType may someday make them a viable alternative.

One, as of yet experimental but promising, change in the technology has a few developers looking at the possibility of connecting extremely small LCD panels directly to the microprocessor. They present the potential of lower cost than the direct-view panels now used in notebooks and FPDs for the desktop PC. Some of the more common technologies being employed to create miniature screens include the Active Matrix Liquid Crystal Display, Field Emissive Displays, Active Matrix Electro-Luminescent displays, and Ferroelectric Liquid Crystal displays.

Figure 3.2. Rocket eBook holds about ten books at one time.

Courtesy of NuvoMedia, Inc. Source: www.rocket-ebook.com/epsshots.

In a dramatic application of the trend toward miniaturization, Xybernaut has incorporated an LCD about 1 inch square into the design of its wearable PC, the Mobile Assistant IV. This Pentium-based 233-megahertz, 32-megabyte RAM, 2-gigabyte hard-drive computer is so small, it fits into a belt that is worn around the waist. An adjustable headset holds the LCD monitor on either side of the head. Jeffrey R. Harrow, a Senior Consulting Engineer at Compaq Computer Corporation and editor of the online journal, *The Rapidly Changing Face of Computing* describes it:

> And on the right side of the headset is THE best head-mounted display I've yet to see. When I was wearing this very light head-mounted display, my right eye simply saw a full-color, full-size(!) 640x480 PC display. It wasn't blurry, too small, or distorted—it was just there, and very readable! It accomplishes this feat with some electronic smoke and mirrors—the bright LCD display points AWAY from you, and the special mirror hanging in front of it redirects and focuses the image into your eye (Harrow 1998).

Microvision, Inc. (http://www.mvis.com), headquartered in Seattle, Washington, is attempting to do better than the Xybernaut solution. The

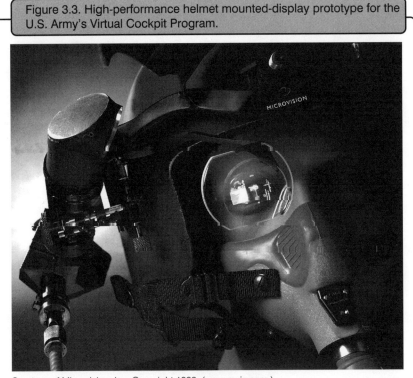

Figure 3.3. High-performance helmet mounted-display prototype for the U.S. Army's Virtual Cockpit Program.

Courtesy of Microvision, Inc. Copyright 1999. (www.mvis.com.)

design of its head mounted visual output apparatus uses no display at all. Instead, its technology, known as Virtual Retinal Display (VRD), uses the human eye's own mechanisms for picking up visual data. The company's publication explains the process:

> The device conveys the image by scanning an electronically encoded beam of light through the pupil to the retina, stimulating the receptors on the back of the eye. The user has the impression of viewing a high quality video image an arm's length away. With its small size and superior performance characteristics, VRD technology can be incorporated into a variety of small hand-held or head-worn devices, making it ideal for a wide array of personal display applications.

References

Apple Computer. 1999. "Apple, Compaq, Matsushita (Panasonic), Philips, Sony and Toshiba Announce Support for IEEE1394 and Plans to Form Patent Pool." http://www.apple.com/pr/library/1999/feb/17firewire.html. (17 February 1999).

Foley, Mary Jo. 1998. "Sun Envisions Networking at Your Fingertips." *PC Week Online*, 15 July 1998. http://www.zdnet.com/zdnn/stories/news/0,4586,336526,00.html. (4 December 1998).

Gillmor, Dan. 1998. "Revolution Online Just Beginning." *Mercury News,* 19 December, 1998. http://www.mercurycenter.com/columnists/gillmor. (20 December 1998).

Harrow, Jeffrey R. 1998. "The Rapidly Changing Face of Computing," 23 November 1998. http://www.digital.com/rcfoc/981123.htm. (25 November 1998).

Intel. 1997. "Intel Announces a New Class of Flash Memory Products." Intel Corporation, 17 September 1997. www.intel.com/pressroom/archive/releases/FL091797.htm. (10 October 1998).

CHAPTER FOUR
Social Issues

T his chapter is designed to provide a brief look at some of the more interesting controversies, conflicts, and issues that have arisen around the latest innovations in digital technology. Currently and in the recent past, the initiatives of and the attacks against the Microsoft Corporation define much of what takes place on the commercial side of the computer industry. Hence, the recent trials and tribulations of Bill Gates' very successful corporation constitute the first issue presented here.

The next issue covered is one that seems to have been with us since the Internet became a popular medium for communication. It is the continuing debate over how to achieve a balance between an individual's right to privacy and society's need to protect its citizens from both prurient media content and from criminal and terrorist cyber attacks. After brief looks at how Intel's latest technology allows for users' Internet habits to be tracked and the role of encryption in the privacy landscape, the chapter moves to a discussion of particular subgroups in society that are being shortchanged in the transition to the Information Age. These subgroups either lack access to the means of exchange in the new economy, or they have come up against barriers to training.

A MICROSOFT WORLD?

When Massachusetts Institute of Technology economics professor Franklin Fisher took the stand as the last prosecution witness in the Department of Justice's (DOJ) antitrust court case against the Microsoft Corporation, he uttered words that many computer experts have expressed within their own peer groups for years. He was being asked to respond to the possibility that integrating a Web browser application (Internet Explorer) into the Microsoft's Windows 98 operating system would create a simpler approach to accessing the Internet and all the services available there. Professor Fisher had to agree that it would make the task of connecting to the Web easier for the average user, but if it was allowed by the court, "We're going to live in a Microsoft world" (Lawsky 1999).

The U.S. government and 19 individual states filed suit against Microsoft, the world's largest software maker, because they have concluded that the company had a functional monopoly in personal computer operating systems and that it was using its dominant position to take control of emerging opportunities on the Internet. Even though Apple Computer distributes its machines with the highly regarded Mac operating system (OS), IBM had provided an alternative graphical user interface called OS2 for years, and Linux and the BeOS are being loaded more often as of late, some estimates place distribution of Microsoft operating systems (MS-DOS, Windows 3.1, Windows 95, Windows 98, and Windows NT) at over 90% of the market. The first thing almost every PC owner sees when they boot up is a desktop owned by Microsoft. The DOJ maintained that it leveraged this advantage by tying its Internet Explorer browser to the OS in order to thwart competition from Netscape Communications, Inc. and Sun Microsystems' Java.

The theory is that since users will have easy, direct access to the Internet Explorer icon (representing the preloaded browser application), Microsoft has a giant head start in the effort to become the most often utilized onramp to the Information Superhighway. The Netscape Navigator products were on the scene months before Microsoft decided that the Internet was truly an important aspect of the computing experience. Critics have contended, and the U.S. government has argued, that in order to gain market share for its own late entry into the browser wars, Internet Explorer (IE) debuted as a free product. Netscape Navigator commercial releases were not free, so the contrast immediately cost Netscape potential revenue. Microsoft countered that this was only normal business practice.

Microsoft also contends that IE is a superior product in many ways, and that by bundling the application with the operating system, they were involved in the best traditions of innovation to benefit consumers. But an internal Microsoft presentation was shown at the trial that disputed this claim. In it, the point was made that the best way to assure the triumph of Microsoft's Internet Explorer over Netscape's Navigator was to make sure IE was installed on new computers before consumers had a chance to decide.

On November 5, 1999, Judge Thomas Penfield Jackson issued findings of fact in the government's antitrust lawsuit against Microsoft. For the most part, Judge Jackson found for the plaintiff, the United States government, and determined that Microsoft did wield monopoly power over the industry. The final paragraph of these findings are included here.

> Most harmful of all is the message that Microsoft's actions have conveyed to every enterprise with the potential to innovate in the computer industry. Through its conduct toward Netscape, IBM, Compaq, Intel, and others, Microsoft has demonstrated that it will use its prodigious market power and immense profits to harm any firm that insists on pursuing initiatives that could intensify competition against one of Microsoft's core products. Microsoft's past success in hurting such companies and stifling innovation deters investment in technologies and businesses that exhibit the potential to threaten Microsoft. The ultimate result is that some innovations that would truly benefit consumers never occur for the sole reason that they do not coincide with Microsoft's self-interest (United District Court for the District of Columbia 1999).

Appeals on the decision may continue for years, but the case has served to expose some of Microsoft's aggressive licensing practices and marketing techniques that may have hindered competition and innovation in the industry. As early as 1996, the company seems to have felt a need to defend its position as the dominant software company in the world. As *Computer Dealer News* reported that year

> The software industry has always shown a strong ambivalence towards Windows. On the one hand, people seem paranoid about the prospect of Windows Global Domination (WGD) and its corollary, The World According to Microsoft. At the same time, developers know a successful platform when they see one.
>
> But then the Internet happened. During the past 18 months, there's been a conceptual shift from the Internet as a cool tool for e-mail and such, to the Internet as computing's future. While corporate IT managers have long championed client/server computing, the Internet has accelerated interest in it. Everyone seems to have discovered—or rediscovered-network-centric computing.

If this is truly the beginning of the end of WGD, there'll be one event that history marks as the time we really knew the winds shifted: the Internet World show in San Jose, CA (April 30 to May 1, 1996). According to reports coming out of that show, a new computing reality emerged—Java (Tanaka 1996).

The Threat to Microsoft

By 1999 Java had matured to the point where it began to live up to the hype that had accompanied its release in 1995. For many programmers and users, Java appeared to be the best possible alternative to the Microsoft juggernaut that had been in place since 1983. Sun Microsystems, better known for the servers they made, developed the new programming language Java (and then the simpler Java script) to take advantage of the change in computing being driven by the Internet. Sun, under the direction of CEO Scott McNeally, foresaw a time when applications would reside, not on the individual's hard drive, but on powerful servers that one would connect to over a network like the Internet. Java was designed as an open system that could run on any computer platform's microprocessor: Sun, Intel, PowerPC, Silicon Graphics, etc. Java was thus a threat to Microsoft, which relied on selling regular upgrades to it's operating system software.

In response, Microsoft pushed it's own ActiveX controls, an API (application programming interface) that added multimedia functionality and Java-like features to Web pages. The race, as each camp knew, was to become the de facto standard for enhancing Internet applications. In the beginning, only Netscape included Java implementation, and it never used ActiveX controls. When Microsoft eventually licensed Java to include its functions in IE, but it broke its agreement to maintain "100% pure Java" coding so that sometimes applets would only operate under a Windows environment. Sun actually sued Microsoft to win a court ordered injunction that forced the company to live up to its licensing agreement.

Unbowed, Microsoft continued to insist that the agreement allowed licensed developers to implement Java in a way that would enhance their own products. Sun, in a move designed to add even more functionality to their product and to increase acceptance, made the code for their latest implementation of Java available to developers via a hybrid type of open source agreement. This new "Java 2" deal, as it is now known, allows anyone to download the development kit for free, but to pay a licensing fee as independent products are shipped. Sun will make certain that compatibility with published specifications for Java are maintained by licensees.

Sun Microsystems also agreed to work with America Online when it purchased Netscape Communications at the end of 1998, again trying to position Java as the ubiquitous alternative to Microsoft's operating systems and Internet browser application. It is unclear whether Sun's McNeally has helped to cut some of the dominance of Bill Gates and his company, but then he isn't alone in that effort.

Larry Ellison's Obsession

Besides the legions of independent programmers who relish the opportunity to improve Java, Netscape, and Linux code, almost as a duty to make alternatives to Microsoft available, no one should underestimate the determination of Larry Ellison, the outspoken CEO of Oracle, the world's second largest software company.

One of Ellison's key initiatives to gain market share for his newly released Oracle8*i*, an Internet-centric database tool that serves up information on most of the top commercial sites of the network, is to make that software work on any platform. His direct competition from Microsoft is the product called SQL Server 7.0, an integrated software package that facilitates a natural language query interface for relational databases. The Gates's plan is to deploy SQL Server 7.0 on thousands of smaller servers connected to Local Area Networks (LANs). In contrast, Ellison's Oracle8*i* is optimized to serve up huge databases at very high response rate with many more supported simultaneous connections than SQL Server 7.0 is reported to be able to support.

The difference in approach does not end here. Ellison claims that Microsoft's operating system, Windows NT, actually slows his software down. For that reason, and because his strategy calls for creating roadblocks to Gates wherever possible, Oracle has been ported to both Unix and Linux operating systems, and it integrates Java as well. More importantly for the future, Oracle has announced the availability of an "Internet appliance" in conjunction with hardware vendor Hewlett-Packard (HP). Making good on a claim that Oracle does not even need an operating system to operate, Ellison announced that HP "is the first computer manufacturer expected to build the Oracle8*i* Appliance previously code-named Raw Iron. HP intends to manufacture and sell computers with preinstalled and preconfigured Oracle8*i* Appliance software, eliminating the need for a stand-alone operating system. The software is expected to dramatically lower the cost and complexity of computing and significantly improve performance for customers" (Oracle 1999).

One thing is certain, as long as Microsoft works as it has to maintain its advantages in the industry, there will be many competitors who work equally hard to take it down. The future, as defined by the realities of Internet computing, will be an interesting one.

PRIVACY

The loss of privacy that results from the application of the new digital telecommunications technologies has been the center of a thorny debate since use of business computing became standard operating procedure. When one considers just how much data is routinely collected about an individual's life: marital status, health history, credit record, employment experience, educational achievement, telephone numbers and telephone call records, banking activities, magazine subscriptions, library records, e-mail addresses and posts to newsgroups, and what sites are visited on the Web, it isn't surprising that public interest groups have been formed to defend against abusive use of this data.

The Electronic Frontier Foundation (EFF) and the Electronic Privacy Information Center (EPIC) are two of the most active organizations monitoring how online and digital information is being used by corporations and government agencies. They, along with other consortia, have advocated the adoption of a privacy statute for the United States that has the same force as similar legislation adopted by the European Union in 1995. From a practical business point of view, failure to take this issue seriously has the potential of creating a situation where a trade war might erupt across the Atlantic Ocean. In late 1998 the European Union passed a tough new law that forbids transmitting personal data to any nation that does not guarantee comparable privacy protections. Economic considerations aside, privacy advocates make the point that a solid foundation for a democratic society relies on the ability of the electorate to act and communicate independently of outside monitoring.

An Intel Threat to Privacy

In January 1999, EPIC led an effort against the latest technological advance in microprocessors announced by the world's leading chip maker, Intel. EPIC explained the situation this way on their special Web page (www.bigbrotherinside.com):

> Intel announced on January 20 that it was planning to include a unique Processor Serial Number (PSN) in every one of its new Pentium III chips. According to Intel, the PSN will be used to identify users in electronic commerce and other net-based applications.
>
> We believe that providing a unique PSN, which can be read remotely by Web sites and other programs in mass-market com-

puters would significantly damage consumer privacy. This number is designed to be used to link user's activities on the Internet for marketing and other purposes (Big Brother Inside 1999).

According to Intel vice president Patrick Gelsinger, the PSN will be used to identify users for accessing Internet Web sites or chat rooms. He told the RSA conference that unless users are able to deliver the processor serial number, they are not able to enter that protected chat room. According to Intel, the technology will also be used for authentication in e-commerce, which will make the PSN attached to a person's real-world identity.

Figure 4.1. Michael Vatis, Director of the National Infrastructure Protection Center, a law enforcement agency working with the FBI, announces the FBI's investigation into the so-called Melissa virus.

Reprinted with permission of Reuters/Blake Sell/ Archive Photos.

In defending the PSN feature of its chip, Intel officials have said that they are relying on the industry to self-police so that abuses of privacy, like the sharing of consumer buying and Web surfing patterns, will be mitigated. EPIC and others were not satisfied. An announced software patch that could be used to turn off the PSN feature, was not an adequate response according to EPIC. They decided to initiate a boycott of Intel to put pressure on them to permanently disable the PSN. Intel has reported that many developers have been in favor of the feature.

Strong Encryption Controversy

The technological tool most often cited by the EFF, EPIC, and others that would aid individuals in the effort to keep their most sensitive information private while online is strong encryption of data. They have fought a battle against the U.S. government, which has called for an escrowed key encryption scheme. This solution that would allow the Federal Bureau of Investigation (FBI) and another designated government agency to hold keys that could decipher any messages transmitted using the approved encryption algorithm. Terrorist threats are most often cited as a reason for the FBI to hold the "wiretap" key, as it has been labeled. No determination has yet been made as to the form of encryption that will become the standard for online activity, but the government has forbid export of strong encryption technology developed by firms like RSA Data Security, Inc.

The securing of data for transport by government agencies and financial institutions around the world is generally achieved using the U.S. government's Data Encryption Standard (DES) algorithm. This current 56-bit standard offers little protection from dedicated hackers. In order to prove that point, the RSA DES Challenge has been run every year since 1997, when RSA Data Security sponsored the contest to crack the government code. The first year, it took 96 days to break the DES algorithm. In 1999, the feat was accomplished in just 22 hours and 15

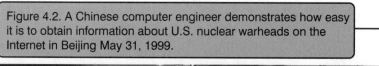

Figure 4.2. A Chinese computer engineer demonstrates how easy it is to obtain information about U.S. nuclear warheads on the Internet in Beijing May 31, 1999.

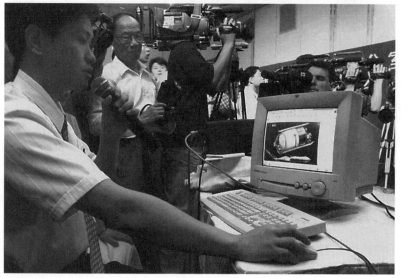

Reprinted with permission of Reuters/Andrew Wong/Archive Photos.

minutes using a mainframe from EFF and over 100,000 personal computers distributed along the Internet.

Free Speech versus Protection

In 1996, the Congress attempted to pass a law that would restrict certain types of information (text, video, graphics, and anything else) that was pornographic in nature. The stated intent of the Communications Decency Act (CDA) an amendment to the Telecommunications Act of 1996, was to protect children from the more prurient interests that seemed to have proliferated over the Internet since it became a commercial venue. After a well organized "blue ribbon" protest campaign and a lawsuit were initiated by a consortium of organizations and corporate partners to protect First Amendment rights on the Internet, the courts eventually struck down the statute. The justices noted that the law was too broadly defined, and would restrict free speech more narrowly on the new medium than what was required on the old.

The Clinton administration and Congress have continued to look for ways to monitor and control sexual (and to a lesser extent, bigoted and hate-based) content on the Net. An attempt was made to resurrect CDA once more in 1998, when the Senate approved legislation that would then be reconciled with a House bill being considered in 1999. This activity continues even though the House of Representatives own Web server (www.house.gov/house/Starr.htm) was responsible for publishing what many have called "the salacious details" of the Clinton-Lewinsky matter then under investigation by the House Judiciary Committee.

The inconsistent nature of the government's approach to regulating the Internet is also evident in the matter of business activity over the network. President Clinton's interagency task force on electronic commerce led by Ira Magaziner published the report "Framework for Electronic Global Commerce" in 1997. That study called for a hands-off approach by government as it related to the e-commerce then beginning to grow on the Web. Noting that, "Self-regulation in the digital age will require the private sector to engage in much greater collective action to set and enforce rules than was characteristic of the Industrial Age" (Clausing 1998), Magaziner and his staff were fearful that government regulation would get in the way of developing private standards for digital transactions. Congress for its part, has voted not to impose transaction taxes on Internet sales for the near term.

The e-commerce issue will be monitored by lawmakers and the public interest groups, especially as that segment of the economy continues to grow at the rapid pace demonstrated over the past three years. Will the government feel compelled to eventually tax Internet sales? And if it does, how will it collect and distribute tariff revenue from a transaction that originates from a buyer on a computer in Hawaii placing an order for goods displayed on a server in Switzerland that represents a company in Oklahoma whose main distribution center is Taiwan? More study is needed.

THE HAVES AND HAVE-NOTS

One of the most discussed sections of the Telecommunications Act of 1996 was the portion on expanding the concept of universal access that was a longstanding policy of standard telephone service in the United States for decades. This is the provision in the old telecommunications law that required telephone companies, originally the Bell Telephone monopoly, to subsidize telephone service in unprofitable areas like rural and remote districts by charging more to urban and business customers. The government at the time saw that a ubiquitous communication system was in the best interest of the country. The universal service provision was a useful solution that achieved the goal. Virtually every household in the United States has a phone.

When the latest telecommunication law went into effect, it was designed to increase competition among various segments of the industry (cable, local and long distance telephone, wireless), and to lower barriers to mergers and cooperative ventures. Everyone was aware that to built the next generation communications system, private concerns would need to operate with less restrictions on how they raised funds and interoperated with their competition. In exchange for increasing profit opportunities, the act acknowledged that social needs must also be met

> to make available, so far as possible, to all the people of the United States without discrimination on the basis of race, color, religion, national origin, or sex a rapid, efficient, Nationwide, and world-wide wire and radio communication service with adequate facilities at reasonable charges. . .(Telecommunications Act of 1966).

At that time, Federal Communications Commission Chairman Reed Hundt called the universal service provision one of the three most important accomplishments of the Act. He said, "Nothing could be more inspiring than the vision of major progress in the global fight against poverty, disease, and misery. Nothing less than that is at stake in our effort to spark sustained, significant, competition-driven growth in our communications and information sector, as ordered by Congress in the landmark Telecommunications Act of 1996. This is about opportunity for everyone" (Hundt 1996). The plan was to provide for access to

Internet types of services, not by buying the less fortunate computers and modems, but by establishing a discount for network wiring and Internet services for institutions where the public traditionally gathered for information: schools and libraries. The e-rate, as it is known, was to be a 20% (approximately) discount to community-based educational centers. More than 30,000 institutions have applied for what Reed anticipated to be about $2.25 billion in subsidies.

By 1999, the e-rate's future was in jeopardy. Because long-distance companies were tapped to pay for this discount out of their profits, it was agreed during negotiations around the Telecom bill, that they would get a reduction in the access charges they paid to local phone companies. They have recently claimed that this break is insufficient, even when the number of schools and libraries receiving aid was cut and the gross amount estimated to cover the subsidy was lowered to $1.275 billion. For that reason, they intend to raise their own rates to their customers for long-distance access. Because some in Congress have viewed this raise in rates as a de facto tax on the public, they have authored bills to eliminate the e-rate plan altogether.

The plan was initiated to help minorities and the poor, people who are well underrepresented among computer owners and Internet users, maintain limited parity with their more well-to-do neighbors. The issue may become more important in a few years when more government services, shopping, education, and even medical access move to the Internet or another type of digital service. Without access to the currency of the Information Age, fast, accurate data, the gap between those who have succeeded and those who have limited opportunities may widen rapidly and precipitously.

WOMEN IN INFORMATION TECHNOLOGY

Even though women constitute approximately 50% of the work force in the United States, their gender is substantially underrepresented in job categories that relate to all areas of science. And where they are employed in scientific fields, they are paid less than their male counterparts. Furthermore, fewer women pursue higher education in the sciences than men. A recent survey sponsored by the National Center for Education Statistics (NCES), "Findings from The Condition of Education 1997: Women in Mathematics and Science" states in its summary conclusion

> Women have made important advances in education over the last few decades, closing the gender gap in the level of educational attainment among younger women that existed 20 years ago. In fact, for several years, women have been awarded the majority of

associate's, bachelor's, and master's degrees. However, a gender gap still exists with respect to mathematics and science, and it widens as students climb the education ladder. Although boys and girls have similar mathematics and science proficiencies at age 9, a gap begins to appear at age 13. At age 17, there is some evidence that the gender gap in mathematics and science has narrowed over time, although a substantial gap remains.

While women are just as likely as men to go to college immediately after high school, from the start they are less interested in majoring in mathematics and science. Although women tend to major in different subjects than men in college, some of these differences have narrowed over time. The mathematics and science fields continue to be areas where the gender gap remains large. Women are far less likely than men to earn bachelor's degrees in computer science, engineering, physical sciences, or mathematics (NCES 1998).

There are many reasons why girls and women do not choose to follow the paths that lead to careers in computer science and the entire range of technically-oriented disciplines. These are well-documented and include societal perceptions and pressures regarding the role of women in society (sexism), girls' own interests and expectations about how human beings should relate to technology.

Marianne Crew, the CEO of Technology Service Solutions, offers an answer to the problem of too few females in the information technology (IT) workplace. She believes that it will take girls being exposed to engineering, math, and science in the early grades so that they can become as familiar and comfortable with that environment as boys are today. The number of women in computer fields is growing, but the challenge of reaching parity with men in compensation and representation will continue for the near term.

ADAPTIVE TECHNOLOGIES

As computer engineers toil to improve the human-machine interface, the ease with which one can use computers to solve problems, tell tales, or communicate with other people has made great strides. For people with disabilities, these more powerful and more intuitive devices have sometimes meant the difference between staying shut away from the world or engaging it on a level rarely experienced in earlier ages. For example, the new text-to-speech and voice-recognition software engines developed by IBM, Dragon, and others have proven their usefulness in the commercial market in the past couple of years. For the visually-impaired or for the person with limited use of hands, this computer input-output solution has proven to be of obvious value. And as increasingly more powerful microprocessors have become less expensive, other adaptive (or assistive) technologies (AT) have been more widely distributed.

Apple Computer was the first company in the industry to create a division specifically oriented to adaptive design consideration. Since 1985, the Disability Solutions Group has provided input to the company regarding universal access features on Macintosh computers, and it has monitored third party products that can provide assistive interfaces for particular needs. Other computer designers have implemented similar initiatives so that special keyboards, new types of switches, pointing devices, etc., have become more commonplace.

But one area of computing, while seeming to improve the ability to get a message out to a wider audience, has actually thrown up an unintended roadblock to people who can't see well. As the Web has evolved into an interactive, multimedia experience, it has left behind many users who once were able to navigate cyberspace using text-based browsers like the freely available Lynx application. Once the new standards like dynamic HTML, cascading style sheets, Java enhancements, and animated images were widely incorporated in Web page design, sightless individuals were cut out of the communication.

The DO-IT project (Disabilities, Opportunities, Internetworking & Technology) is an adaptive technology program located at the University of Washington under the direction of Sheryl Burgstahler, Ph.D. Primarily funded by the National Science Foundation to recruit students with disabilities into science, engineering, and mathematics academic programs and careers, DO-IT makes extensive use of adaptive computer and science technology and the Internet. The program won the President's Award in Mentoring, the National Information Infrastructure Award in Education, and was featured in the 1997 President's Summit on Volunteerism. Burgstahler has pointed out the importance of making Web pages accessible to all visitors, especially as this form of media makes greater inroads against the traditional forms of media.

Burgstahler notes that some people accessing a Web page may have trouble understanding what is being presented for one of many reasons, for example

- They cannot see graphics because of visual impairments.
- They cannot hear audio because of hearing impairments.
- They have slow connections and modems or older equipment that cannot download large files.
- They have difficulty navigating sites that are poorly organized with unclear directions because they have learning disabilities,

speak English as a second language, or are younger than the average user.
- They use adaptive technology with their computer to access the Web.

The DO-IT recommendations for Web page designers who want to make the necessary adaptations that will assure greater accessibility for all visitors are delineated at the DO-IT site (weber.u.washington.edu/~doit/index.html). The following are key requirements that should be met:

- Maintain a simple, consistent page layout throughout your site.
- Keep backgrounds simple. Make sure there is enough contrast.
- Use standard HTML.
- Design large buttons.
- Include a note about accessibility.
- Include short, descriptive ALT tags in html code for all graphical features on your page.
- Include menu alternatives for image maps (also called ISMAPS) to ensure that the embedded links are accessible.
- Include descriptive captions for pictures and transcriptions of manuscript images.
- Caption video and transcribe other audio.
- Make links descriptive so that they are understood out of context.

References

"Big Brother Inside." 1999. Big Brother Inside Homepage. Electronic Piracy Information Center, Junk Busters, and Privacy International. http://www.bigbrotherinside.com/. (17 December 1998).

Clausing, Jeri. 1998. "Internet Commerce Study Stresses Self-Regulation." *New York Times,* 30 November 1998.

"DO-IT." 1999. Project Do-It, University of Washington. http://weber.u.washington.edu/~doit/index.html. (25 January 1999).

Hundt, Reed. 1996. Statement of Chairman Reed E. Hundt, Federal Communications Commission. www.fcc.gov/speeches/Hundt/hundtacs.html. (22 August 1998).

Lardner, James. 1999. "Corporate Evel: Larry Ellison's Next Daring Stunt: Taking on Microsoft." *U.S. News Online,* 18 January 1999. http://www.usnews.com/usnews/issue/990118/18orac.htm. (3 August 1998).

Lawsky, David. 1999. "Trial Witness Warns of a 'Microsoft World.'" Reuters Limited, 7 January 1999. http://www.excite.com. (15 January 1999).

NCES. 1998. *Findings from the Condition of Education 1997: Women in Mathematics and Science.* National Center for Education Statistics. http://nces.ed.gov/pubs97/97982.html. (16 August 1998).

Oracle. 1999. Press release, 25 January 1999: "HP First to Build Oracle8*i*, Appliances." http://www.oracle.com/cgi-bin/press/. (2 February 1999).

Tanaka, David. 1996. "Java and the End of WGD." *Computer Dealer News,* Vol. 12, 30 May 1996, p.14(1).

United States District Court for the District of Columbia. 1999. Findings of Fact in the case of the United States of America vs. Microsoft Corporation, 4 November 1999. http://usvms.gpo.gov/findfact.html. (15 December 1999).

United States Telecommunications Act 1996. http://www.itu.int/sites/wwwfiles/act.txt. (12 February 2000).

CHAPTER FIVE
Biographical Sketches

The individuals profiled in this chapter have been instrumental in advancing computer science or the business of information technology through the years. The time in which their contributions occurred may not be the period which is the direct focus of this book. However, their work has proven to be so significant that it has either directly resulted in changes that were documented in the last four chapters, or it has laid an important foundation for the advances apparent today. These biographical sketches are presented in alphabetic order by last name.

Marc Andreessen, (1972–)

Andreessen has revolutionized the face of computing in recent history. Still in his twenties, the programmer from a small town in midwest America, educated at the University of Illinois, is now one of the fastest success stories of the Internet era. A product of working-class parents and New Lisbon, Wisconsin, public schools, Andreessen did not even touch a computer until he reached the sixth grade. But since computers were one of the things he found most interesting to study, he taught himself how to program in Basic by reading a borrowed book. When his school library finally installed a PC, the first program he wrote was a calculator that helped him do math. Andreessen was an excellent student, scoring well in all subjects. As a reward, and to encourage his interest in computers, his parents got him a Commodore 64 computer when he was 13.

Reports concerning his personality have noted his habit of jumping from subject to subject, as if his brain contained a series of hyperlinks to other dimensions. Once he enrolled at the University of Illinois at Urbana-Champaign, this characteristic proved to be an advantage as he threw himself into computer studies. He was very interested in the Internet, which at that time was seeing a rapid spurt in growth as Tim Berners-Lee had just introduced the Web protocol. But the fact was that the Internet was still most useful only for those who were used to its arcane Unix command structure, which supported e-mail, newsgroups, file transfer, and gopher protocols for archiving, sending, and retrieving data. The hyperlinks of the Web helped in this regard by placing an organizing structure under the millions of files linked via Internet provider (IP) protocol, but it was not yet a mature medium of exchange.

In November 1992, Andreessen thought of a way to improve the Web's functionality. He proposed his solution to Eric Bina, a graduate of the University of Illinois then employed by the National Center for Supercomputing Applications (NCSA). Andreessen was still an undergraduate, but he had found work at NCSA producing code at minimum wage. His idea, however, was anything but small. What if, he asked Bina, all of the functions of the Web (file transfers, viewing, etc.) could be hidden inside a graphical user interface that provided seamless access to all the data? Bina agreed to help write a program to do just that.

Most of the programming work was done under the leadership of Bina who was far more skilled in this department. Andreessen acted more as project manager. He would recruit other programmers, for instance, to port the original Unix code to Windows and Macintosh. The work on the first browser, which they named NCSA Mosaic, was done in six weeks. The first demonstration of the software was in January of 1993. It was a simple program, but it was tremendously effective in opening up the Web to people who had no experience with the Internet in the past. During that first year of its release, millions of copies were downloaded from the NCSA servers.

Since both Bina and Andreessen were employees of NCSA, they could not hope to capitalize on the success of the program. When he graduated from the university in December 1993, Andreessen went to California to work as a programmer for Enterprise Integration Technologies. Soon thereafter he received an e-mail from Jim Clark, a longtime computer industry executive who eventually would suggest that they form a company together. Their Netscape Communications was recently acquired by America Online, and it controls about half of the browser market worldwide. Andreessen is a millionaire many times over.

John Vincent Atanasoff (1903–1995)

Born on October 4, 1903, in Hamilton, New York, Atanasoff created the ABC computer. His device predated the John Mauchy's ENIAC by several years, which most people considered to be the first electronic digital computer ever made. And though the results of a 1972 lawsuit to contradict this assumption have not been widely distributed, the legal system has declared that John Mauchly may very well have borrowed many of his concepts from Atanasoff's groundbreaking ABC machine.

Atanasoff was very interested in mathematics and the principles behind the way numbers worked as a child. His mother gave him a college algebra text, which he was able to complete while still in grade school. After his family relocated to Florida, Atanasoff completed high school in two years. He earned a bachelor of science degree in electrical engineering from the University of Florida, receiving straight A's. The scholar went on to earn a master's degree in mathematics from Iowa State College and a Ph.D. in physics from the University of Wisconsin. By 1936, he was an associate professor of mathematics and physics at Iowa State College.

Once settled into his teaching career at Iowa State, Dr. Atanasoff could afford the time to do some research into the subject that had interested him for many years, facilitating a better method for calculating the results of numerical operations. Not satisfied with the numerous analog calculators then available on the market, Atanasoff reasoned that a digital device would be faster and perform more accurately than the IBM tabulator then in widespread use.

He developed four assumptions that would guide his subsequent work in the design of his new computing machine. These included the use of electricity as the medium of exchange for data, the implementation of a base-two numbering system (binary) as the language, the creation of memory modules based on electric condensers, and the utilization of direct logical action (versus enumeration then used in analog devices) to bring about results.

Clifford Berry, a graduate student in electrical engineering at Iowa State, agreed to help Dr. Atanasoff bring his prototype to life. By December 1939, they were ready to demonstrate the first working model of their Atanasoff Berry Computer (ABC). The machine's potential was proven, and Atanasoff decided to patent the processes he and Berry had implemented in the ABC.

Soon afterward, Atanasoff was at a lecture given by Dr. John W. Mauchly where he introduced himself to the speaker. After some discussion about the ABC, Mauchly asked if he could observe the operation of the device himself. Atanasoff agreed. Five years later the ENIAC was built at the University of Pennsylvania by Mauchly and others, using many of the concepts first demonstrated in the ABC. After a long legal process to determine if intellectual piracy had taken place regarding the ABC, U.S. District Judge Earl R. Larson ruled in 1972 that the ENIAC was indeed derived from the ideas of Atanasoff. Although Judge Larson did not explicitly say that Mauchly "stole" Atanasoff's ideas, Judge Larson did say that Mauchly had used many ideas of Atanasoff's.

Tim Berners-Lee (1955–)

He is called the "Father of the Web" for his work that began at CERN (European Laboratory for Particle Physics) in Geneva, Switzerland. Initially employed at the international energy research center as a consulting software engineer in the second half of 1980, Berners-Lee was looking for a more efficient way to collaborate and share information with physicists. As he puts it, "it was something I needed in my work." His first attempt at a program that would facilitate this goal was software he called Enquire. While never published or used outside of CERN, the program was the basis for a hypertext linking system that he would revisit 10 years later.

Born and educated in England, Berners-Lee received his degree at Queen's College at Oxford in 1976. The computer scientist worked at several jobs in the high-tech sector in the United Kingdom, including two years with telecom equipment manufacturer Plessey Telecommunications and two more years at D.G. Nash where he wrote software that included a typesetting program for intelligent printers and a multitasking operating system. He moved to CERN during six months of the one and one-half years he spent as an independent consultant. In 1981, he returned to his homeland to take on employment with John Poole's Image Computer Systems where he remained until 1984.

It was that year that he took the opportunity of a fellowship at CERN to work on distributed real-time systems for scientific data acquisition and system control. In the simplest of terms, all he wanted to accomplish was the ability to share charts, graphs, and data with other scientists. Berners-Lee, like all computer scientists, knew of the work done on hypertext in the past. The hypertext concept first appeared in a 1945 article by Vannevar Bush, *As We May Think*, in which he proposed Memex, a data-linking system within an intelligent machine. Twenty years later, Ted Nelson coined the terms hypertext and hypermedia in a paper to the Association of Computing Machines of 1965. There he defined hypertext as "nonsequential writing—text that branches and allows choice to the reader, best read at an interactive screen." The first true hypertext application was to come two years after the Nelson paper, when a team led by Dr. Andries van Dam at Brown University created the Hypertext Editing System, which was sold to the Houston Manned Spacecraft Center to produce the documentation for the Apollo space program.

Other work on hypertext would follow, but all of the work prior to Berners-Lee contribution centered on retrieval and mapping of data located on one machine. Dr. Berners-Lee's innovation was to apply the idea to a network of connected computers. As the Internet was just beginning its first era of rapid expansion, creating a program to more readily navigate this new information space was a welcome tool. He wrote the first Web server and the first client (a "browser") in the NeXTStep operating environment. The first implementation of the program was only for internal CERN use in 1990. It was named "WorldWideWeb." Berners-Lee then expanded the work he had done on the CERN application and made it freely available on the Internet in the summer of 1991.

Berners-Lee could have spun off a corporation to license and control the Web protocol and the environment it created. Had he done so, one might argue that the explosion in growth of the Web that accelerated at such an unbelievable rate throughout the decade of the 1990s would have never occurred. Instead, Berners-Lee decided to leave CERN to establish the World Wide Web Consortium at Massachusetts Institute of Technology (MIT) in 1994. He is the first director of the nonprofit organization that numbers over 100 member institutions who have joined in development of Web software and standards.

Vannevar Bush (1890–1974)

Bush is widely credited with developing the concept of hypertext, the linking protocol that was later adapted by Tim Berners-Lee for his distributed computing environment built on Internet protocols known as the World Wide Web. In an article published in the July 1945 issue of *The Atlantic Monthly,* Bush predicted the development of a Memex or a "device in which an individual stores all his books, records, and communications, and which is mechanized so that it may be consulted with exceeding speed and flexibility.

Bush was born in Massachusetts in 1890. He attended Medford, Massachusetts' Tufts University, from which he graduated in 1913. Bush taught at that school until 1917. He went on to become an electrical engineer after earning a doctorate from Harvard and the Massachusetts Institute of Technology (MIT). He served as dean of the engineering faculty at MIT after graduation and up to the onset of World War II.

During his tenure at MIT, Bush directed work on several innovative machines, the most important of which was the differential analyzer. This machine, first conceived by the nineteenth century mathematician Charles Babbage as the Difference Engine, was one of the first analog computing devices.

Also during his time at MIT, Bush was instrumental in founding the electronics manufacturer Raytheon Company. He left MIT in 1939, and from then until 1955, Bush served as president of the Carnegie Institution. And even though today he is probably best known and associated with the Memex concept and hypertext, his most important contribution to the future of scientific discovery was made while he served as the Director of the Office of Scientific Research and Development (OSRD), starting in 1941. During that time, Bush helped set the precedent for the federal government's funding of scientific research at the nation's universities. Up until 1945, over 6,000 scientists were involved in work to improve the tools of war. Then, as the war was ending, Bush used his reputation and leadership to redirect America's substantial engineering and science research efforts away from the building of better weaponry. The OSRD would eventually be responsible for the establishment of the National Science Foundation in 1950. Its influence over the scientific community is still one of the key factors in research for all disciplines.

Esther Dyson (1951–)

A perennial on many of the annual "most influential people in the computer industry" lists, Dyson's most recent accomplishment was being elected as the interim Chair of the Board of the Internet Corporation for Assigned Names and Numbers (ICANN). This influential group will soon be the body responsible for assigning IP addresses and domain names to computers connected to the Internet. From the onset of the Internet, this authority has been the jurisdiction of Jon Postel as director of the Internet Assigned Numbers Authority (IANA), working under a contract with the U.S. government. Just before his death in 1998, Postel had instigated a transition from federal oversight to control by an international body that might better represent the new demographics of the emerging broadband network. Dyson's selection as first chief of the fledgling body is a logical choice, given her long history of involvement with developing Internet business and culture.

Chosen by *Vanity Fair* as one of the "50 Leaders of the Information Age," Dyson is respected as one of the most knowledgeable and influential professionals documenting and directing the sweeping changes taking place in this new era. She has also served as a member of the National Information Infrastructure Advisory Council, chairing the Information

Privacy and Intellectual Property subcommittee, is currently a member of the President's Export Council Subcommittee on Encryption, and continues on the board of the Electronic Frontier Foundation, where she recently served as president.

Dyson graduated from Harvard in 1972 with a bachelor's in economics, but her critical experience was as a journalist for the college's newspaper, *The Harvard Crimson*. A job with *Forbes Magazine* as a reporter from 1974 until 1977 taught her a great deal about business, which took her to Wall Street and work as a securities analyst until 1982.

Edventure Holdings is the company she started when she left Wall Street. It is a consulting firm that specializes in providing information and market analysis about emerging and established information technology (IT) businesses all over the world. Edventure Holdings does, however, emphasize its special expertise about eastern-European markets, and especially Russian markets. Dyson speaks fluent Russian and appears as a speaker on regularly scheduled occasions and at all of the important IT conferences in that country.

Her company also sponsors PC Forum. This important computer industry gathering has brought the computer leaders together every year since 1978. By way of her speeches, consultations, her newsletter *RELEASE 1.0*, and her recent book *Release 2.0: A Design for Living in the Digital Age*, Dyson has assured her place as one of the visionaries of the digital age.

Douglas C. Engelbart (1925–)

Engelbart is known for his many innovations in computing technology. Most famous among these is his work on the first mouse device for data input, but this achievement is merely the tip of the intellectual and innovative iceberg. Engelbart's studies in electrical engineering began in 1942 at Oregon State University but were interrupted by the onset of World War II. After serving two years in the Philippine Islands as a radar technician, he was able to return to get his bachelor's degree in 1948. His first work afterwards was with the National Advisory Committee for Aeronautics (NACA) Ames Laboratory, a research center in San Francisco, which would eventually evolve to become the National Aeronautics and Space Administration (NASA).

After three years, he left to learn more about the emerging computer machines then making news in scientific literature. Because of his experience with radar displays, he posited that it was likely that data from a computer could be made available on a similarly configured tube. Even though the closest computer was located in Maryland at the time, he decided to go to the graduate program in electrical engineering at the

University of California at Berkley. He came out of Berkley with his Ph.D. in 1955. While at the school, he also did work on bi-stable gaseous plasma digital devices for which he earned several patents. After teaching at the school for about a year, he moved on to Stanford Research Institute, later to be known as SRI.

The research position at SRI allowed him to expand his work on computing components, and he was awarded dozens of patents in the first two years for work on magnetic adaptations, digital-device phenomena, and miniaturization. In 1959, he began the work that was to define his mission for the long term. He made the assumption that society was speeding up and that the complexity and the urgency with which organizations must seek solutions was increasing at an incredible rate. Not only did the organization have to get better at doing what it had set out to do, but it must find better and faster ways to support that core mission. His proposed solution to this challenge was a digital augmentation of human intellect; the development of tools and systems that would aid the development of future work modes. His concept was known as bootstrapping.

Funds were found in 1963 to start the Augmentation Research Center. Here, in his own research environment, Engelbart was free to develop the type of digital tools and technologies underlying his bootstrapping augmentation idea. For the next two decades, the center rolled out a series of programs that would support the collaboration processes necessary for the sharing of knowledge and the enhancing of human intellect. One of the most important tools to come out of this work was NLS (oNLine System), a hypermedia linking system that was used for collaborative software engineering, technology transfer, and community support activities. When Defense Advanced Research Projects Agency (ARPA) made the decision to network research labs to promote resource sharing, Engelbart saw this as an opportunity to extend NLS to a widely distributed collaboration environment. His lab became the second site connected to Arpanet. He used the NLS to build a directory of resources that was called the Network Information Center (NIC), which he directed until 1977.

Though NLS, which would eventually become Augment after it ended up as a commercial application of the McDonnell Douglas Corporation, has continued to have widespread applicability as a collaborative environment for knowledge workers in various settings, the work that Engelbart led in his center introduced the world to many computer tools now taken for granted. An incomplete list includes, the mouse, 2-dimensional display editing, in-file object addressing, linking, hypermedia, multiple windows, integrated hypermedia e-mail, hypermedia publishing, shared-screen teleconferencing, and distributed client-server architecture.

First Programmers Club see Betty Snyder Holberton

Bill Gates (William Henry Gates III) (1955–)

Gates is best known as the richest man in America, having attained billionaire status at an earlier age (31) than any other person in history. His wealth is the result of his vast holdings in the first software company founded to create programs for the personal computer. Today the Microsoft Corporation that he started with boyhood friend Paul Allen is the largest software company in the world.

Gates was born in Seattle, Washington, in 1955. He showed an early interest in mathematics, business, and computing. To support these propensities, his influential parents enrolled him at the private Lakeside School at the age of 13, which gave him access to some exceptional educational resources. He proved to be an unusual person even at Lakeside, where it is said, he and his friends studied *Fortune* magazine for fun. He also began programming computers at this time, taking advantage of an ASR-33 Teletype (purchased by a parents' group at the school) which was connected to a General Electric (GE) computer over a phone line. In short order, Gates, Paul Allen, Ric Weiland, and Kent Evans had mastered the operation of BASIC, which that computer operated on. The four boys became the Lakeside Programming Group and learned FORTRAN (Formula Translator), LISP (List Processing Language), and PDP-10 machine language too.

The Lakeside Programming Group would eventually write a Common Business Oriented Language (COBOL) payroll program for a computer company in Portland, Oregon, in exchange for free computer time. The group went on to complete the payroll software, and they leveraged their computer time to do additional jobs for which they were paid in cash. Lakeside School asked them to do a scheduling system , and later they created "Traf-O-Data" for the Washington State Road Department.

By 1974, Gates had left Lakeside for the campus of Harvard University. But he dropped out at age 19 to pursue his real loves, programming and making money. He rejoined one of his high school partners, Allen, and they founded their company, Microsoft. The young entrepreneur's first project was an adaptation of the BASIC computer language, then used only on mainframe computers. Gates and Allen predicted, accurately, that there would be a demand for an operating system to run the new hobbyist computer kits that were beginning to proliferate. While their adaptation was an auspicious beginning, the real breakthrough came in 1980, when they were able to license an operating system to International Business Machines (IBM), which was the largest manufacturer of business-oriented devices in the world. IBM was beginning to experiment with the concept of a new "personal computer" machine, which they intended to market the next year. Convinced that the profits were in their core business, selling hardware, they were happy to pay a royalty to Microsoft for every copy of their new MS-DOS (disk operating system) that was included with their IBM PCs. In 1984, MS-DOS *was* the operating system for virtually every personal computer sold. And Microsoft was on the road to dominating the brand new software industry.

Today, the story of Gates and Microsoft is a part of America's history. The company continues to control the market in PC operating systems with Windows 95, 98, and NT.

Betty Snyder Holberton (1917–) and the "First Programmers Club"

A member of the First Programmers Club, Holberton has become famous as one of the group of six women responsible for the painstaking work of programming the world's first successful commercial computer, the ENIAC, at the University of Pennsylvania during World War II.

Even though Betty Snyder was always an exceptional math student, and she enjoyed the discipline, on her first day of class at the University of Pennsylvania, her mathematics professor asked her if she wouldn't be better off at home raising children. After all, there was a war going on in Europe and Japan, and Betty had a duty to the country.

Undaunted, she continued her studies in math, but decided to major in journalism. This was one of the few fields open to women as a career in the 1940s, and it allowed her to study any subject in which she found an interest. One opportunity that presented itself dovetailed into her love for math. The U.S. Army was looking for people who were good with figures to compute ballistics trajectories at their headquarters in the Moore School located on the University of Pennsylvania campus. And because there were very few men available at the time, and since the work seemed to be clerical in nature anyway, Snyder and 79 other female mathematicians were hired to do it.

In the same building where the women were toiling each day on the trajectory calculations, John Presper Eckert and John Mauchly were directing the top-secret work that was creating the first large-scale electronic computer, ENIAC. When the giant 80-foot-long machine was finished, the engineers looked around for someone to program it to accomplish complicated tasks like figuring ballistics trajectories. No one had programmed a computer before, and it seemed to Eckert and Mauchly that the mathematicians working downstairs might have some insight they could add to the development of the process. Betty Snyder (later Holberton) joined 5 other women from the "computor" pool (Kathleen McNulty Mauchly Antonelli, Jean Jennings Bartik, Marlyn Wescoff Meltzer, Frances Bilas Spence, and Ruth Lichterman Teitelbaum) to become the first programmers club and to make ENIAC a functional piece of information technology.

No one at this point in the history of computers used the word *programming* to mean writing binary code. What it required was physically routing electronic pulses through the maze of 3,000 switches and 18,000 vacuum tubes by way of dozens of extended cables. The women, who were classified as subprofessionals by the government that was relying on them, were not allowed into the large room with ENIAC in the beginning. They were considered security risks. They were forced to work with wiring diagrams and blueprints only. But they did learn how the machine was built, what each relay and tube could accomplish; and finally they were allowed into the room to start testing their procedures. In the early months of 1946, they completed a first run of their program

Figure 5.1. ENIAC Computer.

U.S. Army Photo. Source: "Historic Computer Images." http://titp.arl.mil/~mike/comphist/.

for ballistic trajectories. It ran without a hitch, cutting the time for a single calculation done by hand from 20 hours to 30 seconds.

Until recently, the six women's names have not been mentioned in connection with this seminal event in modern computing history. Part of the culture of the time was to keep women out of any limelight, which might be construed as supporting their work away from the home. And in a very real sense, the hardware was the news then. ENIAC and the male engineers responsible for its construction were the stars of the day. Until Bill Gates demonstrated the economic value of software, programming was simply a clerical extra.

Ruth Teitelbaum, Frances Spence, and Kathleen Antonelli followed ENIAC when it moved to more permanent headquarters at the Army's Aberdeen Proving Grounds in Maryland in December of 1946. They taught other women how to program, which was now considered acceptable women's work. Antonelli eventually married John Mauchly, and the other two dropped out of the workforce. Holberton and Bartik went to New York with Mauchly and Eckart to work on their UNIVAC project. Jean Bartik designed the logic for the UNIVAC I. She retired to raise a family, but later returned to the IT business to work on the PDP-8 minicomputers.

Betty Holberton never left her career. Instead she became a member of the COBOL programming language committee and helped to write standards for FORTRAN. She remained a part of the standards development body until 1983.

Grace Murray Hopper (1906–1992)

"Amazing Grace" Hopper was a remarkably productive individual, scoring unmitigated successes in several areas of endeavor throughout her lifetime. Her work in computerized data automation places her in that elite group of digital pioneers responsible for developing one of the most important steps on the path to more powerful calculating machines. But she was also productive in her academic pursuits, and in military service to her country during World War II and beyond where she attained the rank of Rear Admiral.

A very brief overview of her life begins with graduation in 1928 from Vassar College with honors that include Phi Beta Kappa. She went on to earn an M.A. and a Ph.D. in mathematics from Yale University. She returned to Vassar to teach mathematics and received a Faculty Fellowship grant that allowed her to continue taking classes at New York University in 1941. But as the war raged in Europe and the Pacific, Hopper made a decision to contribute to the winning effort. She joined the U. S. Naval Reserve (USNR) and received her training at the USNR Midshipman School in 1943. Her first assignment after earning the commission of Lieutenant JG was at the Bureau of Ordinance Computation Project back at Harvard, where she learned the art of computer programming on the first large-scale digital computer, the Mark I.

She stepped down from active duty following the end of the war, but she continued at Harvard where she taught and did research on the Navy's Mark I and Mark II computers. She was also recruited by the Eckert-Mauchly Computer Corp. in Philadelphia where she helped program the UNIVAC I, which was the first large-scale commercial computer.

When Hopper learned how to program, all operations were written in machine language, the binary code of 0's and 1's. This time-consuming and tedious procedure was prone to produce software code that was full of bugs, and often would not work as planned. The process was so mistaken-ridden that Hopper knew it was imperative to find a new process to automate much of the repetitive tasks in the structure. In 1952, she wrote a revolutionary program that facilitated the first automatic programming of computer language, i.e., the first "compiler." Using Hopper's software, it was possible for the compiler to insert archived code strings each time a common task was called for in a program. This work served as the basis for another one of her inventions, the Common Business Oriented Language (COBOL). This natural language business software programming tool is still in widespread use today.

Steven Jobs (1955–)

Steven Jobs created the world's first successful personal computer, the Macintosh. He showed little aptitude for any particular career throughout his formative years in Mountain View, California, or later when his family moved to nearby Palo Alto. But he was growing up in an area of fruit trees and orchards that would one day be called Silicon Valley. In high school, he took an interest in technological innovation, and he would attend occasional lectures at the Hewlett-Packard plant located a few miles from the family home. He eventually found summer employment with Hewlett-Packard, where he became friends with another one of the workers, Steve Wozniak.

Woz, as he was known, was five years older than Jobs, and he had recently dropped out of classes at the University of California at Berkeley. The two Steves proved to be a good blending of talent. While Wozniak was consumed with making gadgets and reengineering electronic devices, Jobs was a natural salesman with an innate business sense. At the time, Woz was perfecting his "blue box" design, an illegal electronic contrap-

Figure 5.2. Apple Computer CEO Steve Jobs gives the keynote address at Apple's 1999 Worldwide Developers Conference.

Reprinted with permission of Reuters/Lou Dematteis/Archive Photos.

tion that allowed the user to tap into the telephone company system to make free long-distance calls. Jobs thought that the devices could prove to be a money maker, and he helped his partner sell a few of the devices to friends and others around the area.

By 1972 Jobs was at Reed College in Portland, Oregon, but his academic tenure there was short lived. He dropped out of regular classes to pursue a Bohemian lifestyle. Jobs stayed around campus, dropped in on a few classes, and generally dropped out of the mainstream. But by early 1974, he reunited with Woz as they both took jobs with Atari, Inc. as video game designers.

During this time period, Wozniak was holding meetings of his Homebrew Computer Club at his house. He and the other members were most interested in tinkering with gadgets and experimenting with some of the build-it-yourself computer kits then marketed for hobbyists. Jobs,

never as accomplished as Woz in engineering, wasn't interested in experimenting with the electronics. Instead, his goal was the creation of a
personal computer that could be marketed to the masses. He convinced
his friend to help him build one.

Working on designs in Job's bedroom and constructing in the family's
garage, the two developed the Apple I computer in 1976. It was the first
machine to feature onboard ROM (Read Only Memory) that was programmed to tell the machine how to read programs from an external
source, and it had a video interface that hooked into the user's television
set. List price was $666.00 Jobs sold the first 25 machines to the Byte
Shop, an electronics outfit in Mountain View. To raise capital for the
parts to build units to fill the order, Jobs sold his Volkswagon van and
Wozniak sold his Hewlett-Packard scientific calculator.

Wozniak quit his job at Hewlett-Packard to run the research and
development sector of a new company they formed together. It was
called Apple Corporation because that was Job's favorite fruit, and he'd
spent an agreeable summer as an apple picker in an Oregon orchard. In
1977, the company got backing from venture capitalists of $600,000.00
and a new president in Mike Markkula, a former Intel marketing manager. The Apple II was released that year. It was the first mass market
personal computer.

Eventually, Apple would release the Macintosh in 1983, an all-in-one
machine for the general public. The design, which featured a mouse and
a graphical user interface, and the marketing campaign proved to be a
great success. Jobs recruited John Sculley, former head of Pepsi Cola to
take the position of CEO at Apple the next year, and in a power struggle
among the board, Jobs was asked to leave the company. While his pride
was wounded, he left Apple a very rich man in 1985. Woz also left the
company in that year.

Jobs was back in 1986 when he found investors to start another
computer company, which he called NeXT. The goal was to create the
next generation of powerful hardware and software for the sophisticated
user. Although sales of the expensive machines never took off and
hardware development was finally ended, NeXT software proved to be
very popular among developers. When an opportunity presented itself to
buy Pixar computer animation studios from George Lucas in 1986, Jobs
got it for less than $10 million. This was an area where his powerful
operating system might prove very useful.

When Pixar released the world's first fully computer-animated feature
film, Toy Story, in 1995 for Walt Disney Pictures, it became the highest
domestic grossing film of the year. Jobs seemed to be back on top of his
game and at the forefront of a new direction in the digital information

arena once more. At the same time, the company he had founded as a purveyor of the NeXT product, had fallen on hard times, loosing market share and profitability because of some marketing blunders and production miscues. Jobs contacted them about the possibility of using his NeXT software as the basis for the introduction of a new modern OS for the Mac. Apple decided to buy NeXT and asked Jobs to return as a nonsalaried adviser for CEO Gilbert F. Amelio.

As revenue continued its precipitous drop, the pundits started predicting the demise of the embattled Apple computer company. In 1997, the board took a surprising action. It fired Amelio and named Jobs interim CEO, and in January 2000, Jobs announced he was the permanent CEO. Within a year of his taking back the reigns, Apple Computer was back in the black. During that time he led development on the most talked about computer product of the end of the decade, the groundbreaking iMac. Since then, he has helped Apple shed some of its unprofitable divisions and generally won high praise for turning the company around.

Alan Kay (1940–)

Kay is responsible for several innovations in programming with the concept of "object orientation" heading the list. He was born in Springfield, Massachusetts, to an Australian father who designed prosthetic limbs and a mother who was an artist and musician. An avid reader and dreamer, Kay would often design inventive solutions to problems by taking analogies from music, art, and the creative side of life.

Kay displayed a brilliant mind early in his life, learning to read by age three, for example, but he had a great deal of trouble accepting the rules and procedures of public school. He was often disciplined, suspended, or expelled during his school years. He ended up doing a volunteer stint in the United States Air Force, where he took an aptitude test that showed he would make a good computer programmer. The Air Force was working with a popular computer of the day, the IBM 1401, and as he has related, computer programming at the time was considered a low-status position. Women had that job. But he took to it after determining that programming took a certain elegance and style to be done well. During this time, noting that in one instance some programmer had had the foresight to attach procedures for using the data that was to be shared between different runs of the program on separate machines, Kay reasoned that it would prove very beneficial if all programs could run without having to know how the data was represented.

When he left the Air Force he pursued new found interests at the University of Colorado, where he majored in mathematics and molecular biology. He had also done some song writing and performing while in

Boulder, so after graduation in 1966, he flirted with several career options including music, philosophy, and medicine. In the end, he decided on computer science. The University of Utah was the logical destination for Kay, since Colorado had no such program. At the time, the Utah department had only six students and three teachers who operated within the school of electrical engineering.

Once there, Kay began analyzing the basic building blocks of two recently introduced programs, Sketchpad by Ivan Sutherland and Simula by two Norwegian programmers. He was looking for a common denominator that would allow for more elegant programming. After earning his doctorate in computer science in 1969, Kay moved on to Stanford where he dabbled in work in artificial intelligence, but found the effort unsatisfying because of his biology background. He could not see that the work being done in this field was going to lead to anything that resembled true human intelligence. Instead, he followed another interest, the development of a child-size computing device that would look something like a book.

To support this machine, Kay began work on a new programming language in 1971. In it, the developer used what he finally came to determine was a useful analogy for the basic building block of a computer program, the biological cell. These autonomous entities could communicate with each other by passing messages between the cells. Data, a return address, a receiver address, and the operation the receiver was to perform would all be an integral part of each message passed. Kay called this concept object orientation. Smalltalk, the name for this new language, became the model for programming languages that would follow. No longer were designers relegated to building new code for each separate problem they needed to solve. With object orientation, the programmer can use previously written segments or classes that can be given new attributes. This new procedure has set the standard for software development in the last decade.

Gordon Moore (1928–)

Moore is probably most famous as the originator of Moore's Law, which was first described in a 1965 issue of *Electronics* magazine. Then, as a 37-year old computer scientist, he predicted that the complexity of the silicon microchip would double every 12 months. He later amended the term to 18 months, and that timeframe has proven to be uncannily accurate. Because of his insight and leadership in the manufacture of computer microprocessors, Moore has a reputation as one of the most influential people involved in the computer industry and has been held in high esteem since the earliest days of the integrated circuit.

He graduated in 1954 with a doctorate from the California Institute of Technology, expecting to take up an engineering position on some college faculty. The unassuming scholar had planned a life of research and teaching in academia. The job market, however, would not cooperate with his dreams. Finding no good prospects in electrical engineering departments at any of the west coast colleges, he decided to try his hand at weapons design with John Hopkins University; however, he was very eager to enlist with William Shockley, the co-inventor of the transistor, in 1956. Moore joined Schockley and the design team at his new, Shockley Semiconductor. The company only lasted about a year.

In 1957, Moore took what he had learned at Shockley Semiconductor and started Fairchild Semiconductor with fellow Shockley employee, Robert Noyce, and six others from the defunct corporation. The first product that the company designed was the integrated circuit. Moore, himself, led the development of this groundbreaking technology that placed many connectors and functions onto a single insulating surface. Moore and Noyce left Fairchild with the technology and looked for backing to begin production of the new product. They were able to raise $2.5 million with their new partner, Arthur Rock. The company they formed was known as Intel (INTegrated Electronics).

Intel's initial release was a static memory chip in 1968. These semiconductor memories were at least 10 times more expensive than the industry standard at that time, which was the magnetic core memory. Because of this difference in price, continued research was required to increase the power of the chips. In 1970, a breakthrough was made when Intel introduced the first DRAM chip. This dynamic random access memory module was only 1 kilobyte in size, but it boasted the largest capacity of anything ever made previously. A year later, the company struck again with the first microprocessor. That chip contained 2,300 transistors and was cheap enough and powerful enough to act as the central processing unit in intelligent machines. At this point, manufacturers began to abandon the magnetic core architecture, and Intel's place in history was sealed. Moore no longer has an active role at Intel, but his influence is still felt in the company which still sets the standard for microchip production.

Jon Postel (1943–1998)

Postel was one of the original "fathers" of the Internet. In an obituary that was distributed across the Internet on October 17, 1998, Postel was remembered by his friend and colleague, Vint Cerf:

Out of the chaos of new ideas for communication, the experiments, the tentative designs, and crucible of testing, there emerged a cornucopia of networks. Beginning with the ARPANET, an endless stream of networks evolved, and ultimately were interlinked to become the Internet. Someone had to keep track of all the protocols, the identifiers, networks, addresses, and ultimately the names of all the things in the networked universe. And someone had to keep track of all the information that erupted with volcanic force from the intensity of the debates and discussions and endless invention that has continued unabated for 30 years. That someone was Jonathan B. Postel, our Internet Assigned Numbers Authority, friend, engineer, confidant, leader, icon, and now, first of the giants to depart from our midst (Cerf 1998).

Postel was born in Altadena, California, in 1943 and grew up in the suburbs there. He demonstrated only average abilities through high school but became interested in computer science while attending community college. His transfer to the University of California at Los Angeles (UCLA) to pursue a degree in engineering would prove to be a timely and fortuitous move. This was the time when the government was initiating work on the concept for a ubiquitous communications network that would be capable of maintaining integrity and functionality even if a section were to be eliminated in a nuclear strike or a natural disaster.

Along with Bob Taylor, Larry Roberts, and Vint Cerf, he helped develop principles for connecting computers in what they called a "distributed network." He is credited with installation of this network's (ARPANET) first communications switch to route packet traffic in 1969. Though computer science was not even a course of study offered at UCLA when he entered the institution, he was awarded a doctorate in that discipline in 1974.

Perhaps Postel's most critical contribution to the development of the Internet was his systemization and long-term management of the network's address system. He administered the Internet Assigned Numbers Authority (IANA) for 30 years as it's director. This simple, but important body was responsible for disseminating and tracking IP numbers (domain names) across the Internet. All a computer needs to connect to the network is a unique identification number, or address. This uncomplicated protocol was instrumental in helping to expand the Internet with astounding speed. With Postel at the helm, IANA has kept the Internet a stable, workable communication medium from the days when Arpanet hosted 5 nodes to today when millions of computers access the Web every minute of the day.

Postel was also the editor of the Request for Comments (RFCs) series, which is a system of technical notes and papers that deal with issues related to maintenance of the Internet. The RFCs began with the inception of ARPANET and have evolved to almost 2,500, covering subjects that have eventually become the technical standards of Internet protocol.

Postel became the manager of the networking research division of the Information Sciences Institute (ISI) at the University of Southern California. He was also a founding member of the Internet Architecture Board, the first individual member and a trustee of the Internet Society, the custodian of the U.S. domain, and a founder of the Los Nettos Internet service.

Just prior to his death, Postel shocked much of the Internet's technical community by rerouting the Internet's directory service to alternate locations. He did this to prove that the system in place was robust and redundant enough to display no disruptions in service. At that time, he was also working on the effort to change responsibility for administration of the Internet from IANA to a nonprofit international corporation. That effort is ongoing.

Steve Wozniak (1950–)

Wozniak is best known for his partnership with Steve Jobs in the late 1970s (see Jobs biography), wherein these two imaginative entrepreneurs were able to combine their interest in the emerging technologies to create the first personal computer for the general public. Their creative energies were responsible for Apple Computers, a company started in their garage in 1976 when they designed a $666.00 computer that could hook up to the family television set. That machine featured a ROM chip that would allow it to read and operate software from other sources, and it was a hit mainly among the hobbyist set. Their second generation machine, the Apple II, was even more important, finding a place in small businesses and in school classrooms all over the United States. It featured color output and was supported by many independent programmers who supplied countless software products to enhance its usefulness. It has been referred to as the Volkswagen of computers, and most of the design work was done by Wozniak. By the time the Macintosh appeared in 1983, Apple was the standard for easy-to-use personal computing, and their products were in the marketplace four years ahead of IBM's MS-DOS personal computer.

"The Woz," or "the Wizard of Woz," as he was known, demonstrated a knack for mathematics and engineering that went beyond the norm. It was rumored that his mother would actually have to grab and shake him out of a math-induced reverie. He would start thinking about a problem, and often nothing would distract his effort. He always wanted to be an engineer, and in 1971, he enrolled in the University of California at Berkeley to major in that discipline. But he soon dropped out, choosing to get hands-on experience with one of the foremost information technology firms of the day, Hewlett-Packard. He was employed as an engineer at the Palo Alto site from 1973 to 1976. It was during this time that he joined in the infamous undertaking to make "blue boxes" with John Draper. These small devices allowed a telephone user to make long-distance calls without being charged by the phone company. Before he quit this illegal business, Steve Jobs had entered his life, joining with Woz to sell the devices. Steve Jobs would then move away, going to college and later India for a couple of years.

When Jobs returned to Palo Alto in 1975, he found his friend Wozniak still at Hewlett-Packard, but also leading his own Homebrew Computer Club. This club was full of engineering enthusiasts who were intent on getting their hands on the latest chips, boards, and circuitry, so that they could create new, smaller computing devices. Wozniak was the most knowledgeable of the group, and his design efforts were bringing about a prototype that sparked the interest of Jobs. The computer Wozniak was working on was eventually built and marketed with the help of Jobs and the Apple Computer company they formed became one of the most important IT companies of the age.

Wozniak left Apple in 1981 after his private plane crashed, and he suffered a head injury, which left him with no short-term memory for about five weeks. He decided to take time off, and go back to the things he loved to do. He re-enrolled at Berkley using the name "Rocky Clark," with the intent of finally earning his computer science degree. This he did receive, along with a electrical engineering degree in 1982. He returned to his position as head of research and development at Apple, but soon decided to end his participation all together. Wozniak had decided that his efforts were better suited to the social, not corporate side of life. And after leaving the company he helped found, he became involved in various causes, especially those directed at helping children learn. To this day he holds classes in his own house for children interested in using technology to enhance their lives and those of their neighbors.

References

Cerf, Vint. 1998. "I Remember IANA." RFC 2468. Internet Society, 17 October 1998. www.internet.isi.edu:80/in-notes/rfe/files/frc2468.txt.

CHAPTER SIX
Documents and Speeches

T his chapter contains five documents concerning the development of computer technology and some of the issues that have been generated around its implementation in the last few years. The documents are in order according to the topics covered in the first few chapters of this book.

The first document is the transcript of a speech given by the inventor of the World Wide Web, Tim Berners-Lee, on the occasion of his receipt of the Distinguished Fellowship of the British Computer Society on July 17, 1996, at the new British Library in London. In his talk, Berners-Lee provides insight into the development of the tools that created the Web, and he envisions a roadmap for the future.

The second document is a reproduction of the Memorandum of Understanding between the U.S. Department of Commerce and Internet Corporation for Assigned Names and Numbers. This important draft agreement will be the basis for a new methodology and structure for coordinating the assignment and tracking of Internet Protocol (IP) addresses as the Internet continues its expansion.

To provide a perspective on the controversies inherent in the way the world's most powerful software company markets its products, the third document is the petition filed by attorneys representing the United States Department of Justice as it sought to find Microsoft in contempt of a court order to cease requiring original equipment manufacturers (OEMs) to install "other company products" as a condition of preloading the Windows operating system on their computers. This action was a precursor to the 1998 antitrust suit that sought to significantly alter the way Microsoft does business.

A short excerpt from testimony at that trial is the fourth offering in the chapter. This piece is from a Microsoft defense witness, Professor Richard L. Schmalensee of MIT.

The final document is provided by the Progressive Policy Institute of Washington, D.C., and is included with the permission of one of its authors, Rob Atkinson, director of Technology and New Economy Project for that organization. Along with Marc Strassman, he has written a policy briefing, *Jump-Starting the Digital Economy (with Department of Motor Vehicles-Issued Digital Certificates)*, that argues for the implementation of an authentication system for individuals using the Internet.

THE WORLD WIDE WEB—PAST, PRESENT, AND FUTURE

Tim Berners-Lee

The transcript below is reproduced with the permission of the British Computer Society (www.bcs.org.uk). It is the text of a speech presented by Berners-Lee, inventor of the World Wide Web protocol and a key advocate for maintaining rigorous standards for this powerful communications medium.

It is a great honour to be distinguished by such a Fellowship, and I should immediately say two things:
- One is that, of course, the Web was developed by a whole lot of people across the Internet, who discovered about it on Internet User Groups and went away with the ideas and started playing and encouraging each other, and developing a little grass-roots community. The Web owes an incredible amount to those across the Internet, and also to the "bosses who didn't say no" who are now all wearing halos across the planet, and who really enabled this sort of thing to grow to the point where they didn't have the option of saying no.
- The other thing I would say is that the ideas existed before putting it together in the World Wide Web, which is basically trivial.

So what's special about it? What I think we are celebrating then is the fact that dreams can come true. So many times it would be nice for things to be this way but they don't come out for one reason or another. The fact that it did work is just so nice; that dreams can come true. That's what I've taken away from it, and I hope that it applies to lots of other things in the future.

Figure 6.1. Tim Berners-Lee, inventor of the World Wide Web, gives a speech at CERN in Geneva in 1998.

Reprinted with permission of Pierre Virot/Reuters/Archive Photos.

I'll go back now over a little bit about the origins, a bit about the present, and just a little bit about the future, very much in overview.

The Past

The original intent of the Web was that it should be let's start with a definition the "universe of network accessible information." The point about it being a universe is that there is one space. The most important thing about the Web is this URL space, this nasty thing which starts with *http*. The point of a URL is that you can put anything in there, so the power of a hypertext link is that it can point to absolutely anything.

That is why, whereas hypertext had been very exciting beforehand, and there had been a little community that had been happily going on for several years making hypertext systems that worked across a disc or across the file system, when the Web allowed those hypertext links to point to anything and it suddenly became a critical mass, it became really exciting. Maybe that will happen to some other things as well.

In fact the thing that drove me to do it (which is one of the frequently asked questions I get from the press or whoever) was partly that I needed something to organise myself. I needed to be able to keep track of things, and nothing out there, none of the computer programs that you could get, the spreadsheets and the databases, would really let you make this random association between absolutely anything and absolutely anything, you are always constrained.

For example, if you have a person, they have several properties, and you could link them to a room of their office, and you could link them to a list of documents they have written, but that's it. You can't link them to the car database when you find out what car they own without taking two databases and joining them together and going into a lot of work. So I needed something like that.

I also felt that in an exciting place like CERN, which was a great environment to be in and to start this. You have so many people coming in with great ideas, doing some work, and leaving with no trace of what it is they've done and why they did it the whole organisation really needed this. It needed some place to be able to cement, to put down its organisational knowledge.

And that idea, of a team being able to work together, rather than by sequence of grabbing somebody at coffee hour and bringing somebody else into the conversation, and having a one time conversation that would be forgotten, and a sequence of messages from one person to another.

Being able to work together on a common vision of what it is that we believe that we are doing, and why we think we are doing it, with places to put all the funny little "this is why on Tuesday we decided not to do that." I thought that would be really exciting, I thought that would be a really interesting way of running a team, maybe we could work towards that goal: that dream of the "self-managing team."

So, that was why these were the original goals. Universal access means that you put it on the Web and you can access it from anywhere; it doesn't matter what computer system you are running, is independent of where you are, what platform you are running, or what operating system you've bought and to have this unconstrained topology, which because hypertext is unconstrained it means you can map any existing structures, whether you happen to have trees of information or whatever.

As people have found, it is very easy to make a service which will put information onto the Web which has already got some structure to it, which comes from some big database which you don't want to change, because hypertext is flexible, you can map that structure into it.

In the early days, talking over tea with somebody, I was comparing it to a bobsled: There was a time before it was rushing downhill that there was quite a lot of "pushing" to be done. For the first two years, there was a lot of going around explaining to people why it was a really good idea, and listening to some of the things that people outside the hypertext community come back with.

The hypertext community, of course, knew that hypertext was cool, and why doesn't everybody like it? Why doesn't everybody use it? People felt that hypertext was too confusing—the "oh, we'll be lost in it won't we" syndrome. Also I was proposing to use an SGML type syntax. SGML at the time was mainly used in a mode whereby you would write an SGML file and you would put it in for batch processing perhaps overnight on an IBM mainframe and with a bit of luck you would find in the morning a laser printed document.

But the idea of doing this SGML parsing and generation of something that could be read in real time was thought to be ridiculous. People also felt that HTML was too complex because "you have to put all those angle brackets in." If you're trying to organise information get real you're not going to have people organising it. You "can't ask somebody to write all those angle brackets just because they want to make something available on an information system, this is much too complex."

Then there was also a strong feeling, and a very reasonable feeling at CERN, that "we do high-energy physics here." If you want some special information technology, somebody was bound to have done that already, why don't you go and find it. So that took me with a colleague, the first convert, Robert Cailliau, to the European Conference on Hypertext at Versailles where we did the rounds of trying to persuade them, all these people who had great software and great interfaces and had done all the hard bits to do the easy bit and put it all on-line.

But having, perhaps due to lack of persuasive power, not succeeded in that, it was a question of going home and taking out the NeXT box.

Using NeXT was, I think, both a good and a bad step. The NeXT box is a great development environment, and allowed me to write the WorldWideWeb Program. (At that time it was spelled without any spaces. Now there are spaces, but for those of you who are interested in that sort of thing, there are no hyphens).

So the WorldWideWeb was a program I wrote at the end of 1990 on the NeXT. It was a browser editor and a full client. You could make links, you could browse around it was a demonstration which was fine but, of course, very few people had NeXT and so very few people saw it.

At CERN, there was a certain amount of raised eyebrows and it was clear that we wanted it on MAC, PC and Unix platforms but there wasn't the manpower to do it. So it we went around conferences and said "hey, look at this. If you have a student, please suggest they go away and implement this in a flashier way and on one of the platforms please." There was a couple of years of that.

There was also the Line Mode Browser which was the first real proof of universality. The Line Mode Browser [is] a very simple Web browser that runs on a hard-copy terminal. All you need is the ASCII character set, carriage line feed, and you can browse and print a node out, with little numbers by all the links at the bottom, and you can choose a number. (I mention these things just because sometimes its worth remembering that the path through A to B is sometimes through C, D and E, F, and G in totally different places.)

It was necessary to put the Line Mode Browser out to get other people who didn't have NeXT able to access the Web, so that nobody has the excuse not to be able to access it. The next thing I see is a newspaper article saying that the WorldWideWeb is a system for accessing information "using numbers."

There is a snowball effect here. It is very difficult when you produce a new information system. You go to someone and say "hey, look in here" and they say "What? What have you got?" "Its all about the World Wide Web," they say "big deal." So you say "why don't you put some more information in here" and they say "who's looking in it?" and you have to say "well, nobody yet because you haven't put any information in yet." So you've got to get the snowball going. Now that's happened and you can see the results.

Initially the first thing we put on the Web was the CERN phone book which was already running on the mainframe. We did a little gateway which made the phone book appear in Hypertext with a search facility and so on. For the people at CERN there was a time when WWW was a rather strange phone book program—with a really weird interface! During that time gopher was expanding at an exponential rate and there was a strong feeling that gopher was much easier, because with gopher you didn't have to write these angle-brackets.

But the Web was taking off with distributed enthusiasm; it was the system administrators who were working through the night when it got to 6 o'clock in the morning and they decided that "hey, why bother going home" and they started to read "alt.hypertext" (yes, hypertext an alternative news group, one of those alternative sciences). Alt.hypertext is where you had to discuss this sort of thing. These systems administrators were the people who would only read the alternative news groups and they would pick up the software and play with it. Then by 8 o'clock in the morning you'd have another Web Server running with some new interesting facet and these things would start to be linked together.

There were some twists and turns along the winding road, there was my attempt to explain to people what a good idea URLs were; they were called UDIs at the time, Universal Document Identifiers and then they were called Universal Resource Identifiers, then they were called Uniform Resource Locators (in an attempt to get it through the IETF, I consented that they could be called whatever they liked).

I made the mistake of not explaining much about the Web concepts, so there was a 2-year discussion about what one could use identifiers names and addresses for. It is pretty good if you're into computer science, you know you can talk for any length of time about that kind of thing without necessarily coming to any conclusion.

It's worth saying that I feel a little embarrassed accepting a fellowship when there are people like Pei Wei a very quiet individual who took up the challenge. He read about the World Wide Web on a newsgroup somewhere and had some interesting software of his own; an interpreted language which could be moved across the NET and could talk to a screen.

Basically, he had something very like Java, and as he went ahead and wrote something very much like Hot Java, the language was called "Viola" and the browser was called "ViolaWWW." It didn't take off very quickly because you had to first install "Viola," nobody understood why you should install an interpreter, and then this "WWW" in a Viola library area. You had to be system administrator to do all that stuff, it wasn't obvious. But in fact what he did was really ahead of his time. He actually had Applets running. He had World Wide Web pages with little things doing somersaults and what have you.

Then there was a serious turning point when someone at NCSA brought up a copy of "Viola" and Mark Andreesen and company saw it and thought "Hm, we can do that." Mark Andreesen worked the next 14 nights, or something, and had Mosaic.

One other thing he did was put in images, and after that the rest is more or less history. Nothing had really changed from the Line Mode Browser in that Viola was just a browser, it was not an editor. And the same for Erwise which had preceded it. In fact there is another one called Cello which had been written for the PC which preceded Mosaic, and in each case they wrote a World Wide Web client which is a piece of software which can browse around the Web, but unlike the original program you couldn't actually edit or make links very easily.

I think this was partly because NeXTStep software has got some neat software for making editable text, which is difficult to do [with] a WYSIWYG [What-You-See-Is-What-You-Get] word processor from the ground up. But it is also because when you get to the Browser, you get all excited about it, you get people mailing you and end up having to support it and answer questions about it.

Mark Andreeson found himself deluged by excitement in Mosaic, and still we didn't have anything which could allow people to really write/create links easily with a couple of key strokes, until NaviPress—who's heard of NaviPress—a little Company bought by AOL and now called AOL Press. They are still there, and a number of other editors which actually allow you to go around and make links although still not as intuitively as I would have liked.

So those are some of the steps, there are lots of other ones and many anecdotes, but this was the result as seen from CERN [refers to figure showing straight line growth of use of WWW on CERN server, with vertical axis on an logarithmic scale]. This shows the load on the first WWW server. By current terms its not a very big hit rate. Across the bottom is from July '91 to July '94 and there is a logarithmic scale up the side of total hits per day.

The crosses are week days and the circles are weekends and you can see what happened I call that a straight line you can see that every month, when I looked at the log file it was 10 times the length of the log file for the same month the previous year. There are a couple of dips in August and there are a couple of places where we lost the log information when the server crashed and things.

People say "When did you realise that the Web was going to explode like this?" and "when did it explode?" In fact if you look, there was the time when the geek community realised that this was interesting, and then there was the time when the more established computer science and high energy physics community realised that this was interesting, and then there is when *Time* and

Newsweek realised it was interesting. If you put this on a linear scale, you can pick your scale and look for a date on which you can say it exploded, but in fact there wasn't one. It was a slow bang and it is still going on. It's at the bottom of an 'S' Curve and we are not sure where the top is.

The Present

And then after the bang we are left with the postconceptions (the reverse of preconceptions). One of those was that because the first server served up Unix files, there was an assumption that those things after the *http:* had to be Unix file names. A lot of people felt locked into that, and it was only when Steve Putts put up an interesting server where URLs were really strange but would generate a map of anywhere on the planet to any scale you wanted and with little links to take you to different places and change the scale.

After a few other really interesting servers which had a different sort of information space, the message got through that this is an opaque string and you could do with it what you like. This is a real flexibility point, and it's still the battle to be fought. People try to put into the protocols that a semicolon here in the URL will have a certain significance, and there was a big battle with the people who wrote "Browser" that looked at the ".html" and concluded things about what was inside it wrong! URL is not a file name, it is an opaque string and I hope it will represent all kinds of things.

People kept complaining about URLs changing—well, that was a strange one because URLs don't change, people change them. The reasons people change them are very complex and social (and that gets you back into the whole naming and addressing loop) but there was a feeling for a while that there should be a very simple, quick cure to making a name space, in which you would just be able to name a document and anybody would be able to find it.

After a lot of discussions in the ITF and various fora, it became clear that there was a lot of social questions here, about exactly who would maintain that, and what the significance of it was, and how you would scale it. In fact there is no free lunch, and it is basically impossible.

There was the assumption that, because links were transmitted within the HTML that they had to be stored within HTML files, until people demonstrated that you could generate them, on the fly, from interesting programs. And from the assumption that clients must be browsers it seemed to follow that they can't be editors—for some reason although everybody has got used to WYSIWYG in every other field, they would not put up with WYSIWYG in the Web.

But people had to write HTML you have to write all those angle brackets. It was one of the greatest surprises to me that the community of people putting information on line was prepared to go and write those angle brackets. It still blows my mind away, I'm not prepared to do it, it drives me crazy.

But now we hear back from these people who got so into writing the angle brackets, that HTML is far too simple; we need so many exciting new things to put in it because we need to be able to put footnotes, and frames, and diagonal flashing text that rotates and things. Didn't things change over those few years?

And where are we now? Well, what you actually see when you look at the Web is pretty much a corporate broadcast medium. The largest use of the Web is the corporation making a broadcast message to the consumer. I'd imagined initially that there would be other uses and I talked a bit about group work business, but clearly once you've put something up, if there is any incentive—whether it is psychological or monetary or whatever, because your audience is very large, it is very easy for you to push it up the scale, it pays you very much to go for that global audience. You can afford to put in so much more effort if you have got a global audience for your advertising, for your message, subtle or not. So that is what is seen. And there is some cool stuff.

There is VRML; 3-D sites where you wander through 3-dimensional space, maybe that will become really interesting (actually I think it will happen because to do 3-D on a machine you need a fast processor but you don't need a fast phone line and I think the fast processors are coming a lot faster than the fast phone lines). So 3-D is something which may happen a long time before video.

There are style sheets coming out which will allow you to do that flashing orange diagonal rotating text. You can redo all your company wide documents with the flick of a button just by changing the style sheet without having to change all that HTML.

There's Java, which is really exciting. At last the Web has given the World an excuse to write in a decent programming language instead of C or Fortran. Begging your pardon, there have been object oriented programming languages before now, but if a real programmer programmed in one, typically the boss would come round and say "sorry, that's fine, son, but we don't program like that in this organisation," and you have to go away and re-write it all in C. Just the fact that the Web has been there to enable a new language to become acceptable is something.

What's the situation with the Web itself as an information space? From the time when there was more than one browser, there was tension over fragmentation. Whenever one browser had a feature, an adaptation of the protocol, and the other one didn't, there was a possibility that the other one would adapt, would create that feature but use a slightly different syntax, or a very different syntax, or a deliberately different syntax.

You get places where you find a little message which says, "this page has been written to work with Mosaic 5.6 or NetScape 3.0 or Internet Explorer 2.8 or whatever it is, and its best for you to use that browser."

And now? Do you remember what happened before The Web? Do you remember this business when you wanted to get some information from another computer: you had to go and ask somebody how to use this telnet program, you had to take it to someplace and FTP files back onto a floppy disc, you picked the floppy disc up and went down the corridor, and it wouldn't even fit in your computer!

When you got yourself a disc-drive that would take it, then the disc format was wrong, so you got yourself some software, and with someone's help you could read the format on it, and what you got was a nice binary WordStar document and there was no way you could get it into Word Perfect or Word Plus 2.3—remember that? Do you remember how much time you spent doing all that?

Well, the people who put these little things at the bottom of their Web pages saying this is best viewed using "Fubar' browser are yearning, yearning to get back to exactly that situation. You'll have 17 Web browsers on your page and you'll get to little places which say "now please switch to this" and "now please switch to that" and suddenly there is not one World Wide Web, but there a whole lot of World Wide Webs.

So if any of you have got Web Masters out there put those little buttons on there saying this is best used using a particular browser, suggest they put "this is best used using a browser which works to the following specifications: HTML 3.2," or something like that. You can go back this evening; e-mail them from your homes, and tell them that I just mentioned it.

So there is a tension of fragmentation, what are we going to do about it? In 1992 people came into my office, unannounced, from large companies, sometimes more than one company at a time. I remember one in particular when four people came and sat down around a table and banged it and said "Hey, this Web is very nice but do you realise that we are orienting our entire business model around this. We are re-orienting the company completely, putting enormous numbers of dollars into this, and we understand the specifications are sitting on a disc you have somewhere here. Now what's the story, how do we know it is still going to be there in 10 years, and how do we put our input into it?"

I asked, of course, what they felt would be a good solution to that, and I did a certain amount of touring around and speaking to various institutes, and the result was I felt there was a very strong push for a neutral body. Somewhere where all the technology providers, the content providers, and the users can come together and talk about what they want; where there would be some facilitation to arrive at a common specification for doing things. Otherwise we would be back to the Tower of Babel.

So hence the Web Consortium. The Consortium has 2 hosts; INRIA in France for Europe, and MIT for North America. We are also looking at setting up various things in the Far East. We have 145 members at the last count (maybe it's 150 now it seems that the differential between my counting and my talking about it is 5). We are a neutral forum—we facilitate, we let people come together.

We actually have people on the staff who have been editing Web specs, are aware of the architecture, are basically very good. They can sit in on a meeting and edit a document, know when people are saying silly things, and produce a certain amount of advice. We have to move fast.

We are not a standards organisation, I'm sorry. We do not have meetings from every one of our 150 or whatever it is, countries in the world sitting round, and we do not have 6-month timescales. Sometimes we have to move extremely rapidly when there is a need for something in the marketplace and the community wants to have a common way of doing it. So we don't call what we do "standards" we call them "specifications."

We have just introduced a new policy by which we can simply ask the members whether they think something is a good idea, and if they do then we call it a "recommendation" as opposed to a "standard." In fact what happens is that when we get together, the engineers who know what they are talking about from the major players (they are primary experts in the field), write a little piece of specification, put their names on it, and its all over bar the shouting. Everybody takes that "spec" and runs with it, de-facto "standards" arrive in most cases.

But every area is different and so we have to be very flexible. Some areas we have to consult, we have to be more open, there are more people who want to be involved. In some areas we have to just move extremely rapidly because of political pressure.

At the same time we like to keep an eye on the long-term goals, because although the pressures are fairly short-term there is a long-term architecture. There are some rules in the World Wide Web; like the fact that URLs are opaque; like the fact that you don't have to have *http* at the beginning of a URL but you can move onto something else; like the fact that HTTP and URLs are independent specifications and HTML is independent of HTTP, you can use it to transport all kinds of things.

If, originally, the specs had fixed that the World Wide Web uses HTTP and HTML we wouldn't have Java applications or other things being transported across the Web. We wouldn't be able to think about new protocols.

The Future

Its worth saying a word about the long-term goals. There is still a lot of work before this can be an industrial strength system, so that when you click on the link you know you are going to get something.

There are a lot of things that have got to change, such as redundancy which has got to be able to happen, just fixing everything "under the hood" so that you can just forget about the infrastructure. Something which is very complicated, involves some pretty difficult problems in computer science, and its important.

More on the evident side; I have a horizontal scale between the individual human interaction at the end, through to the corporation talking to the masses? I'd originally imagined that the point about the Web was that you would also be able to have personal diaries, and in that personal diary you'd be able to make a note, and you'd be able to put a pointer to the family photograph album, and your brother's photograph album, which are just accessible to the family, or the extended family.

You would be able to put a pointer to a meeting you've got to go to at work, but the meeting agenda would be just visible to and used by a little group of people working together, and that in turn would be linked to things in the organisation of the town you are living in, such as the school. Imagine that you have a range of things going up through what is called the Intranet (the World Wide Web scaled down for corporate use), to the whole global media thing, and that this would all be one smooth continuum.

I thought it was simple, we just needed to get browser/editors which were good and then we would be able to play. To a certain extent that's true.

When we do have browser editors we'll be able to do a lot more, but there is a lot more that you need. You need to have trust; you need to be able to make sure that other people don't see those photograph albums and what have you. There is a lot of infrastructure that has still to be put together, but I am very interested in the Web being used across that scope.

I'm also interested in these machines that we all have on our desks being actually used to help us. What they are doing at the moment is delivering things for us to read, decisions for us to make and information for us to process. For us to process! Hey, what about these computer things? I thought the idea was they were supposed to do some of the work. At the moment they can't.

They could, in fact, do it when its a database but they haven't a chance on the Web because everything on the Web is written in bright shining pink and green for your average human reader, who can read English (who can read pretty bad English at times), so if you and I have difficulty parsing it, going out and asking a machine to solve the problem is pretty difficult at the moment.

Let's suppose there is a house for sale and you want to buy it. You would like to know that person really owns it. Suppose you don't have a Land Registry, so you go and you find the Title Deeds, which are on the Web, as are all the transfers of ownership going way back. They are there, but it's a lot of work to go back through all of them unless they are put in a form that is actually a semantic statement, some knowledge representation language statement.

Knowledge representation is another thing that people have played with, but it really hasn't taken off in a tremendous way on a local scale. Maybe it is something that, if we can get the architecture right globally, then that would take off too. Then you would be able to simply ask your computer to go out and find an interesting house, the sort of house you like, within your price range, and see if it is really owned by the person who is selling it (or whether in fact they sold off half of the back garden 10 years ago but they hadn't told you). It would be able to go and make all the assumptions, it would be able to figure out in fact whether the documents it reads it ought to believe, by tracing through the digital signatures and what have you.

Those are some long-term goals. They are not things that the press, the consortium, the members, the newsgroups, talk about all the time, but they are things we are trying to keep in the back of our minds.

I'll go through very rapidly the areas that W3C is actually developing or could develop. There are basically 3 areas of work:

The User Interface and Data Formats, the parts of the architecture and protocol which are affected by, and specifically affect, the sort of society that we can build on the Web. The sort of things that are in the user interface area are the continual enhancement of HTML for more and more features, putting different sorts of SGML documents onto the Web, solving the internationalisation problem (or at least trying to take those pieces of the internationalisation problem, type-setting conventions, such as type-setting in different directions and character sets) and trying to take those solutions which exist and show how you can use them in a consistent way on the Web. Style sheets, graphics in 3-dimensions. The PNG format, for example, is a new graphics format to replace the graphics interchange format because its bigger and better, which we have been encouraging (although not doing ourselves). Most of the user interface and data formats work is done in Europe.

There is the whole area of Web protocols in society, security and payment and the question of how parents can prevent their children from seeing material which they don't want them to be viewing until they are old enough. It is this pressure to protect a child, until the age of digital majority, particularly in the United States but also in Germany and various other countries, that has produced the Platform for Internet Content Selection, or PICS system. This is an initiative which has produced specifications which should, I hope, be in software and usable by the end of 1996.

There are other exciting things on the horizon. Such as protocols to actually transfer semantic information about international property rights. Can you imagine taking the licence information on the back of a floppy disc, one of those in such small type that if you blew it up to a readable size it would probably be poster size, and actually trying to code that up into some sort of semantic language I can't, but maybe we can work in that direction. Questions of how to find the demographics of who is looking at your site without infringing the privacy of any individual person.

The third area of Web architecture is looking at the efficiency and integrity of the Web. How do you prevent the problems of dangling links; find out when you have linked to a document which no longer exists; rapidly and painlessly.

How do you get copies of heavily used documents out to as many places as you can, all over the planet, and having done that, how does the person in the arbitrary place find out where the nearest one is? These are part of the unsolvable naming problem.

In general, we are aiming to bring the thing up to industrial strength. We had a workshop about this recently. There is the question of whether these things we find on the Web are really objects and what does that mean? Does this mean that the distributed object world should somehow merge, there should be some mapping between Web objects and distributed objects. What does this mean?

And that raises the question of mobile code objects, which actually move the classes around. There are lots of exciting things going on, not that the average user would notice apart from the fact that they get little gismos turning corners of the tops of their Web pages when Java applications come over.

There is just one more thing that I want to emphasise. I initially talked about the Web, and said that I wanted it to be interactive. I meant this business about everybody playing at the level where you have more than one person involved but not the whole universe. Perhaps you've got a protected space where you can play.

I feel that people ought to be able to make annotations, make links and so get to the point where they are really sharing their knowledge. I talked about interactivity. I found people coming back to me and saying "Isn't it great that the Web is interactive" and I'd say "Huh? "Well you know you can click on these buttons on forms and it sends data right straight into the machine."

I felt that if that is what people meant by interactivity then maybe we need another word (I say this with total apology because I think people who make up new words are horrible) but lets just for the purpose of this slide talk about intercreativity, something where people are building things together, not just interacting with the computer, you are interacting with people and being part of a whole milieu, a mass which is bound together by information.

Hopefully with the computers playing a part in that too. To do that we need to integrate people with real-time video that you hear so much about. Why isn't it better integrated with the Web? Why can't I when I go to the library the virtual library that is find people's faces and actually start talking to them? Why don't I meet somebody in the library?

The nice thing about the virtual library is that you are allowed to talk in it, except that talking protocols haven't been hooked into the Web protocols yet, so we just need to do a little hooking together (Ha! "a little bit of hooking together there" sounds like 3 years work of solid standardisation meetings).

How about having objects that you can manipulate. I'd like to be able to hold a virtual meeting in a 3-dimensional area where there is a table and where you can move the chairs around, and when I move the chairs you see it happen.

We could build graphs and models, mathematical models, real models, engineering models, little models of new libraries, to see if we can make them look nice sitting next to St.Pancras Station or something. I'd like to be able to see all that happen in the Web so that means building into the infrastructure objects; objects which know how to be interacted by many people at once and being able to update their various many instances and copies throughout the World.

The military folks use 3-dimensional digital simulation technology for playing tank battles and maybe there will be some good stuff coming out of that, I don't know. But a very simple thing would be to notify somebody that something has changed. It's great having this model of global information— you write something, I go in and I change it, and put an important little yellow post-it sticker on it, but if you don't find out that I've done it then it's not very much use, so we need to have ways of notifying both people and machines that things have changed.

I would like to see people get more involved in this; at the moment, it doesn't look like one great big television channel, but lots and lots and lots and lots of very shallow television channels and basically the mouse is just a big television clicker. There must be more to life.

So, let me conclude with a few challenges that as a community we have. One is making the most of this flexibility, we have got to, we need to keep it flexible. We need to be able to think our way past the Web as a set of linked Hypertext documents.

Hopefully pretty soon the Web infrastructure, the information space, will be just a given, like we assume IP now (we don't worry about IP, although we should, because its running out of address space and all kinds of stuff and nobody is funding the transatlantic links). We just kind of assume that Internet protocol is there, the Internet is there and we are worrying about how we build the Web on top of it.

We've got to make sure that there is somebody there having the next bright idea and can use that flexibility to make something which has got a totally different topology which is used to solve a totally different problem. To do that, we have got to make sure that we are not, in our designs, constraining the future evolution, we're not putting in those silly little links between specifications.

Let me give you just one example: It is possible with some browsers to put a piece of HTML on the Web. The server delivers it to the browser and inside one of the tags is an attribute, and the attribute value is quoted, and inside the attribute value is a quoted string. It's normally used to be able to write 10 for something like a point size or a width or a whatever but now you can put a little piece of Javascript in there and some browsers, if they don't see 10 but something in curly brackets, they will just send it off to the Javascript interpreter.

Now if you've actually got a Javascript interpreter this is dead easy. You can do that in 2 lines of code; just take the curly brackets off and call Javascript, but just think what's happened. In ten years time, to figure out what it meant, not only do you have to look up the old historical HTML space, but you have also got to find Javascript. Javascript is going to be changing and so you thought you had a nice, well defined language, but it's just one line's reference from that specification to the other specification.

In fact, you've got a whole big language specification except that in one part of it it's got angle brackets, and the other part of it has curly brackets and semi-colons; and they are totally different, one thing is totally incomplete and the other is self modifying. And so, by not saying "by the way this document is in HTML and Javascript 2.0" that one little trap then its the sort of thing which could trip us up later.

The third thing which is really important is that we have to realise that when we define these protocols and the data formats, we are defining things like the topology of the information. We are defining things like who can get access to what information. We are defining things about privacy; about identity; how many identities you can have; whether it is possible to be anonymous; whether it is possible for some central body to do anything at all; whether it is possible for a central body to do lots of things like find out the identity of anonymous people.

Whether there is a right for two people to have a private conversation, which we rather assume at the moment because they can go into the middle of a big field, but does that right hold in Cyberspace? If it does, does this mean that the world will fall apart because terrorism will be so easy? Do all these questions about society come back to the protocols we define, which define the topology in the properties of Cyberspace.

So if you think you're a computer programmer. If you think you're a language designer. If you think you're a techie; and one of the nice things about being a [techie] is that you can forget all that ethics stuff because everybody else is doing that and thank goodness you didn't have to take those courses you are wrong.

Because when you design those protocols you are designing the space in which society will evolve. You are designing, constraining the society which can exist for the next 10-20 years.

I'll leave you with that thought.

MEMORANDUM OF UNDERSTANDING BETWEEN THE U.S. DEPARTMENT OF COMMERCE AND INTERNET CORPORATION FOR ASSIGNED NAMES AND NUMBERS

On November 25, 1998, the U.S. Department of Commerce (DOC) and the Internet Corporation for Assigned Names and Numbers (ICANN) signed an agreement that would give the latter public benefit corporation the responsibility for administering and assigning Internet addresses (the IP numbers and .com, .gov, .net, .edu domain names). This work had been done previously by IANA (Internet Assigned Numbers Authority) under the direction of Internet pioneer, Jon Postel. Much of Postel's work is represented by the organization laid out in the following memorandum. This text is also available at http://www.ntia.doc.gov/ntiahome/domainname/icann-memorandum.htm.

I. Parties

This document constitutes an agreement between the U.S. Department of Commerce (DOC or USG) and the Internet Corporation for Assigned Names and Numbers (ICANN), a not-for-profit corporation.

II. Purpose

A. Background

On July 1, 1997, as part of the Administration's Framework for Global Electronic Commerce, the President directed the Secretary of Commerce to privatize the management of the domain name system (DNS) in a manner that increases competition and facilitates international participation in its management.

On June 5, 1998, the DOC published its Statement of Policy, Management of Internet Names and Addresses, 63 Fed. Reg. 31741(1998) (Statement of Policy). The Statement of Policy addressed the privatization of the technical management of the DNS in a manner that allows for the development of robust competition in the management of Internet names and addresses. In the Statement of Policy, the DOC stated its intent to enter an agreement with a not-for-profit entity to establish a process to transition current U.S. Government management of the DNS to such an entity based on the principles of stability, competition, bottom-up coordination, and representation.

B. Purpose

Before making a transition to private sector DNS management, the DOC requires assurances that the private sector has the capability and resources to assume the important responsibilities related to the technical management of the DNS. To secure these assurances, the Parties will collaborate on this DNS Project. In the DNS Project, the Parties will jointly design, develop, and test the mechanisms, methods, and procedures that should be in place and the steps necessary to transition management responsibility for DNS functions now performed by, or on behalf of, the U.S. Government to a private-sector not-for-profit entity. Once testing is successfully completed, it is contemplated that management of the DNS will be transitioned to the mechanisms, methods, and procedures designed and developed in the DNS Project.

In the DNS Project, the parties will jointly design, develop, and test the mechanisms, methods, and procedures to carry out the following DNS management functions:

    a. Establishment of policy for and direction of the allocation of IP number blocks;

    b. Oversight of the operation of the authoritative root server system;

    c. Oversight of the policy for determining the circumstances under which new top level domains would be added to the root system;

    d. Coordination of the assignment of other Internet technical parameters as needed to maintain universal connectivity on the Internet; and

e. Other activities necessary to coordinate the specified DNS management functions, as agreed by the Parties.

The Parties will jointly design, develop, and test the mechanisms, methods, and procedures that will achieve the transition without disrupting the functional operation of the Internet. The Parties will also prepare a joint DNS Project Report that documents the conclusions of the design, development, and testing.

DOC has determined that this project can be done most effectively with the participation of ICANN. ICANN has a stated purpose to perform the described coordinating functions for Internet names and addresses and is the organization that best demonstrated that it can accommodate the broad and diverse interest groups that make up the Internet community.

C. The Principles

The Parties will abide by the following principles:

1. Stability

This Agreement promotes the stability of the Internet and allows the Parties to plan for a deliberate move from the existing structure to a private-sector structure without disruption to the functioning of the DNS. The Agreement calls for the design, development, and testing of a new management system that will not harm current functional operations.

2. Competition

This Agreement promotes the management of the DNS in a manner that will permit market mechanisms to support competition and consumer choice in the technical management of the DNS. This competition will lower costs, promote innovation, and enhance user choice and satisfaction.

3. Private, Bottom-Up Coordination

This Agreement is intended to result in the design, development, and testing of a private coordinating process that is flexible and able to move rapidly enough to meet the changing needs of the Internet and of Internet users. This Agreement is intended to foster the development of a private sector management system that, as far as possible, reflects a system of bottom-up management.

4. Representation

This Agreement promotes the technical management of the DNS in a manner that reflects the global and functional diversity of Internet users and their needs. This Agreement is intended to promote the design, development, and testing of mechanisms to solicit public input, both domestic and international, into a private-sector decision making process. These mechanisms will promote the flexibility needed to adapt to changes in the composition of the Internet user community and their needs.

III. Authorities

A. DOC has authority to participate in the DNS Project with ICANN under the following authorities:

1. 15 U.S.C. § 1525, the DOC's Joint Project Authority, which provides that the DOC may enter into joint projects with nonprofit, research, or public organizations on matters of mutual interest, the cost of which is equitably apportioned;
2. 15 U.S.C. § 1512, the DOC's authority to foster, promote, and develop foreign and domestic commerce;
3. 47 U.S.C. § 902, which specifically authorizes the National Telecommunications and Information Administration (NTIA) to coordinate the telecommunications activities of the Executive Branch and assist in the formulation of policies and standards for those activities including, but not limited to, considerations of interoperability, privacy, security, spectrum use, and emergency readiness;
4. Presidential Memorandum on Electronic Commerce, 33 Weekly Comp. Presidential Documents 1006 (July 1, 1997), which directs the Secretary of Commerce to transition DNS management to the private sector; and
5. Statement of Policy, Management of Internet Names and Addresses, (63 Fed. Reg. 31741(1998) (Attachment A)), which describes the manner in which the Department of Commerce will transition DNS management to the private sector.

B. ICANN has the authority to participate in the DNS Project, as evidenced in its Articles of Incorporation (Attachment B) and Bylaws (Attachment C). Specifically, ICANN has stated that its business purpose is to

i. coordinate the assignment of Internet technical parameters as needed to maintain universal connectivity on the Internet;
ii. perform and oversee functions related to the coordination of the Internet Protocol (IP) address space;
iii. perform and oversee functions related to the coordination of the Internet domain name system, including the development of policies for determining the circumstances under which new top-level domains are added to the DNS root system;
iv. oversee operation of the authoritative Internet DNS root server system; and
v. engage in any other related lawful activity in furtherance of Items i through iv.

IV. Mutual Interest of the Parties

Both DOC and ICANN have a mutual interest in a transition that ensures that future technical management of the DNS adheres to the principles of stability, competition, coordination, and representation as published in the Statement of Policy. ICANN has declared its commitment to these principles in its

Bylaws. This Agreement is essential for the DOC to ensure continuity and stability in the performance of technical management of the DNS now performed by, or on behalf of, the U.S. Government. Together, the Parties will collaborate on the DNS Project to achieve the transition without disruption.

V. Responsibilities of the Parties
A. General

1. The Parties agree to jointly participate in the DNS Project for the design, development, and testing of the mechanisms, methods and procedures that should be in place for the private sector to manage the functions delineated in the Statement of Policy in a transparent, non-arbitrary, and reasonable manner.
2. The Parties agree that the mechanisms, methods, and procedures developed under the DNS Project will ensure that private-sector technical management of the DNS shall not apply standards, policies, procedures or practices inequitably or single out any particular party for disparate treatment unless justified by substantial and reasonable cause and will ensure sufficient appeal procedures for adversely affected members of the Internet community.
3. Before the termination of this Agreement, the Parties will collaborate on a DNS Project Report that will document ICANN's test of the policies and procedures designed and developed pursuant to this Agreement.
4. The Parties agree to execute the following responsibilities in accordance with the Principles and Purpose of this Agreement as set forth in section II.

B. DOC

The DOC agrees to perform the following activities and provide the following resources in support of the DNS Project:

1. Provide expertise and advice on existing DNS management functions.
2. Provide expertise and advice on methods and administrative procedures for conducting open, public proceedings concerning policies and procedures that address the technical management of the DNS.
3. Identify with ICANN the necessary software, databases, know-how, other equipment, and intellectual property necessary to design, develop, and test methods and procedures of the DNS Project.
4. Participate, as necessary, in the design, development, and testing of the methods and procedures of the DNS Project to ensure continuity including coordination between ICANN and Network Solutions, Inc.
5. Collaborate on a study on the design, development, and testing of a process for making the management of the root server system more robust and secure. This aspect of the DNS Project will address:
    a. Operational requirements of root name servers, including host hardware capacities, operating system and name server software versions, network connectivity, and physical environment.

    b. Examination of the security aspects of the root name server system and review of the number, location, and distribution of root name servers considering the total system performance, robustness, and reliability.

    c. Development of operational procedures for the root server system, including formalization of contractual relationships under which root servers throughout the world are operated.

6. Consult with the international community on aspects of the DNS Project.

7. Provide general oversight of activities conducted pursuant to this Agreement.

8. Maintain oversight of the technical management of DNS functions currently performed either directly, or subject to agreements with the U.S. Government, until such time as further agreement(s) are arranged as necessary, for the private sector to undertake management of specific DNS technical management functions.

C. ICANN

ICANN agrees to perform the following activities and provide the following resources in support of the DNS Project and further agrees to undertake the following activities pursuant to its procedures as set forth in Attachment B (Articles of Incorporation) and Attachment C (By-Laws), as they may be revised from time to time in conformity with the DNS Project:

1. Provide expertise and advice on private sector functions related to technical management of the DNS such as the policy and direction of the allocation of IP number blocks and coordination of the assignment of other Internet technical parameters as needed to maintain universal connectivity on the Internet.

2. Collaborate on the design, development, and testing of procedures by which members of the Internet community adversely affected by decisions that are in conflict with the bylaws of the organization can seek external review of such decisions by a neutral third party.

3. Collaborate on the design, development, and testing of a plan for introduction of competition in domain name registration services, including:

    a. Development of procedures to designate third parties to participate in tests conducted pursuant to this Agreement.

    b. Development of an accreditation procedure for registrars and procedures that subject registrars to consistent requirements designed to promote a stable and robustly competitive DNS, as set forth in the Statement of Policy.

    c. Identification of the software, databases, know-how, intellectual property, and other equipment necessary to implement the plan for competition.

4. Collaborate on written technical procedures for operation of the primary root server including procedures that permit modifications, additions, or deletions to the root zone file.

5. Collaborate on a study and process for making the management of the root server system more robust and secure. This aspect of the Project will address
   a. Operational requirements of root name servers, including host hardware capacities, operating system and name server software versions, network connectivity, and physical environment.
   b. Examination of the security aspects of the root name server system and review of the number, location, and distribution of root name servers considering the total system performance, robustness, and reliability.
   c. Development of operational procedures for the root system, including formalization of contractual relationships under which root servers throughout the world are operated.
6. Collaborate on the design, development, and testing of a process for affected parties to participate in the formulation of policies and procedures that address the technical management of the Internet. This process will include methods for soliciting, evaluating, and responding to comments in the adoption of policies and procedures.
7. Collaborate on the development of additional policies and procedures designed to provide information to the public.
8. Collaborate on the design, development, and testing of appropriate membership mechanisms that foster accountability to and representation of the global and functional diversity of the Internet and its users, within the structure of private-sector DNS management organization.
9. Collaborate on the design, development, and testing of a plan for creating a process that will consider the possible expansion of the number of gTLDs. The designed process should consider and take into account the following:
   a. The potential impact of new gTLDs on the Internet root server system and Internet stability.
   b. The creation and implementation of minimum criteria for new and existing gTLD registries.
   c. Potential consumer benefits/costs associated with establishing a competitive environment for gTLD registries.
   d. Recommendations regarding trademark/domain name policies set forth in the Statement of Policy; recommendations made by the World Intellectual Property Organization (WIPO) concerning: (i) the development of a uniform approach to resolving trademark/domain name disputes involving cyberpiracy; (ii) a process for protecting famous trademarks in the generic top level domains; (iii) the effects of adding new gTLDs and related dispute resolution procedures on trademark and intellectual property holders; and (iv) recommendations made by other independent organizations concerning trademark/domain name issues.

10. Collaborate on other activities as appropriate to fulfill the purpose of this Agreement, as agreed by the Parties.

D. Prohibitions

1. ICANN shall not act as a domain name Registry or Registrar or IP Address Registry in competition with entities affected by the plan developed under this Agreement. Nothing, however, in this Agreement is intended to prevent ICANN or the USG from taking reasonable steps that are necessary to protect the operational stability of the Internet in the event of the financial failure of a Registry or Registrar or other emergency.
2. Neither Party, either in the DNS Project or in any act related to the DNS Project, shall act unjustifiably or arbitrarily to injure particular persons or entities or particular categories of persons or entities.
3. Both Parties shall act in a non-arbitrary and reasonable manner with respect to design, development, and testing of the DNS Project and any other activity related to the DNS Project.

VI. Equitable Apportionment of Costs

The costs of this activity are equitably apportioned, and each party shall bear the costs of its own activities under this Agreement. This Agreement contemplates no transfer of funds between the Parties. Each Party's estimated costs for the first six months of this Agreement are attached hereto. The Parties shall review these estimated costs in light of actual expenditures at the completion of the first six month period and will ensure costs will be equitably apportioned.

VII. Period of Agreement and Modification/Termination

This Agreement will become effective when signed by all parties. The Agreement will terminate on September 30, 2000, but may be amended at any time by mutual agreement of the parties. Either party may terminate this Agreement by providing one hundred twenty (120) days written notice to the other party. In the event this Agreement is terminated, each party shall be solely responsible for the payment of any expenses it has incurred. This Agreement is subject to the availability of funds.

Joe Sims	J. Beckwith Associate Burr
Counsel to ICANN	Administrator, NTIA
Jones, Day, Reavis & Pogue	U.S. Department of Commerce
1450 G Street N.W.	Washington, D.C. 20230
Washington, D.C. 20005-2088	

Parties Estimated Six Month Costs
A. ICANN

Costs to be borne by ICANN over the first six months of this Agreement include: development of Accreditation Guidelines for Registries; review of Technical Specifications for Shared Registries; formation and operation of Government, Root Server, Membership and Independent Review Advisor Committees; advice on formation of and review of applications for recognition by Supporting Organizations; promulgation of conflicts of interest policies; review and adoption of At-Large membership and elections processes and independent review procedures, etc; quarterly regular Board meetings and associated costs (including open forums, travel, staff support and communications infrastructure); travel, administrative support and infrastructure for additional open forums to be determined; internal executive, technical and administrative costs; legal and other professional services; and related other costs. The estimated six month budget (subject to change and refinement over time) is $750,000–$1 million.

B. DOC

Costs to be borne by DOC over the first six months of this Agreement include: maintenance of DNS technical management functions currently performed by, or subject to agreements with, the U.S. Government, expertise and advice on existing DNS management functions; expertise and advice on administrative procedures; examination and review of the security aspects of the Root Server System (including travel and technical expertise); consultations with the international community on aspects of the DNS Project (including travel and communications costs); general oversight of activities conducted pursuant to the Agreement; staff support equal to half-time dedication of 4–5 full time employees, travel, administrative support, communications and related other costs. The estimate six month budget (subject to change and refinement over time) is $250,000–$350,000.

PETITION IN THE UNITED STATES DISTRICT COURT FOR THE DISTRICT OF COLUMBIA

This excerpt from the court document was a follow-up to an initial order from Judge Thomas Penfield Jackson forcing Microsoft Corporation to "unbundle" it's Web browser from the operating system then in development. Immediately following this text is the direct testimony of defense witness, Richard L. Schmalensee Dean of the Massachusetts Institute of Technology Sloan School of Management. His position in the subsequent anti-trust trial, wherein Microsoft was accused of working unfairly to drive Netscape Communications out of the marketplace, is an important counter to the government position. This and other supporting and subsequent documents can be found on the Department of Justice Web page at http://www.usdoj.gov/atr/cases/ms_index.htm.

United States of America, petitioner, v. Microsoft Corporation, respondent.

Supplemental to Civil Action No. 94-1564

Hon. Thomas Penfield Jackson

Petition by the United States for an Order to Show Cause Why Respondent Microsoft Corporation Should Not Be Found in Civil Contempt

Joel I. Klein, Assistant Attorney General, Douglas Melamed, Principal Deputy Assistant Attorney General, Rebecca P. Dick, Director of Civil Non-Merger Enforcement

Christopher S. Crook, Chief, Phillip R. Malone, Steven C. Holtzman, Pauline T. Wan, Karma M. Giulianelli, Michael C. Wilson, Sandy L. Roth: Attorneys, U.S. Department of Justice, Antitrust Division, 450 Golden Gate Ave., Room 10-0101, San Francisco, CA 94102, (415) 436-6660.

The United States of America, by its attorneys, acting under the direction of the Attorney General of the United States, presents this Petition for an order requiring Respondent Microsoft Corporation ("Microsoft") to show cause why it should not be found in civil contempt of the Final Judgment entered by this Court on August 21, 1995 in United States v. Microsoft Corporation , Civil Action No. 94 § 1564 (1994) ("Final Judgment"). A copy of the Final Judgment is attached hereto as Appendix A.

The United States represents to the Court as follows:

Nature of the Action

Microsoft is the world's largest and most powerful personal computer software producer. Through its "Windows" operating system products, it possesses a monopoly in the market for operating system software for Intel© compatible personal computers ("PCs") and from this monopoly enjoys a corporate profit rate and market capitalization that are among the highest of any major American company.

Microsoft unlawfully maintained its monopoly by using exclusionary and anticompetitive contracts to market its PC operating system software. To stop this conduct, the United States sued Microsoft in July 1994 for violating Sections 1 and 2 of the Sherman Act. Microsoft settled that lawsuit by consenting to the Final Judgment, which prohibits Microsoft from imposing various anticompetitive terms in its contracts with PC original equipment manufacturers ("OEMs") that preinstall Microsoft's operating system software products on the computers they sell. Most importantly for this Petition, the Final Judgment prohibits Microsoft from conditioning the terms of an OEM's license to distribute the Windows operating system on the OEM also licensing and distributing other Microsoft products. The purpose of that and other provisions of the Final Judgment was to prevent Microsoft from protecting or extending its operating system monopoly. Microsoft distributes its Windows products to end users primarily through PC OEMs.

Because a PC can perform almost no useful tasks without an operating system, and because shipping PCs without operating systems is likely to cause customer confusion and increase product support costs, OEMs consider it a commercial necessity to preinstall operating system software on nearly all of the PCs they sell. At present, no other operating system software is a commercially viable substitute for Microsoft's Windows 95 operating system. As a result, OEMs overwhelmingly license Windows 95 and preinstall it on virtually all of their PCs.

Microsoft's Windows monopoly gives Microsoft substantial power to force OEMs to license and distribute other Microsoft— software—products by requiring the OEMs to license such products as a condition of receiving their Windows 95 operating system license.

This Petition challenges Microsoft's requirements that OEMs license one such product—Microsoft's Internet browser —along with, and as a condition of, licensing Microsoft's commercially essential Windows 95 operating system. Internet browsers are software products that enable PC users to access, view, and use information and software programs located on the Internet and World Wide Web.

Since August 1995, Microsoft has produced an Internet browser product known as Internet Explorer ("IE"). Microsoft has produced and aggressively marketed several successive versions of Internet Explorer, and over the past year has been promoting the version known as IE 3.0. On September 30, 1997, Microsoft released its newest version of Internet Explorer, known as Internet Explorer 4.0. In direct violation of the Final Judgment, Microsoft has required OEMs, as part of their license for Microsoft's Windows 95 operating system and as a condition of receiving that license, to license, preinstall, and distribute Microsoft's Internet Explorer 3.0 browser on their PCs that also have Windows 95 installed. Microsoft intends to impose similar requirements with respect to Internet Explorer 4.0 beginning on or about February 1, 1998.

Microsoft's conduct—conditioning its Windows licenses on OEMs licensing Internet Explorer—is precisely the sort of improper use of Microsoft's market power to protect and extend its monopoly that this Court's Final Judgment sought to prevent and which it expressly prohibits. Microsoft's current and prospective conditioning constitutes a clear and serious violation of the terms and purpose of the Final Judgment, and requires the Court's intervention.

This violation goes to the heart of the Final Judgment. Internet browser software is not simply another software product. Rather, as detailed below, Internet browsers are an important element in a fundamental competitive challenge that is arising to Microsoft's operating system monopoly. By forcing OEMs to license and distribute Internet Explorer on every PC they ship with Windows 95, Microsoft is not only violating the Final Judgment, but in so doing is seeking to thwart this incipient competition and thereby protect its operating system monopoly. In order to stop Microsoft's ongoing violation and prevent its imminent future violation of the Final Judgment, and thereby protect the integrity and underlying purpose of the Final Judgment, the United States requests that the Court adjudge Microsoft in civil contempt of the Final Judgment and impose the relief requested below.

Jurisdiction of the Court

This Petition alleges violations of the Final Judgment by Microsoft. The Court has jurisdiction over Microsoft under its inherent power to enforce compliance with its orders, pursuant to 18 U.S.C. 401(3) (1988), and under Sections III and VII of the Final Judgment.

Section III of the Final Judgment provides:
This Final Judgment applies to Microsoft and to each of its officers, directors, agents, employees, subsidiaries, successors and assigns; and to all other persons in active concert or participation with any of them who shall have received actual notice of this Final Judgment by personal service or otherwise.

Section VII of the Final Judgment provides: Jurisdiction is retained by this Court over this action and the parties thereto for the purpose of enabling any of the parties thereto to apply to this Court at any time for further orders and directions as may be necessary or appropriate to carry out or construe this Final Judgment, to modify or terminate any of its provisions, to enforce compliance, and to punish violations of its provisions.

Respondent and Its Relevant Products

Microsoft is a corporation organized and existing under the laws of the State of Washington, with its principal place of business located at One Microsoft Way, Redmond, Washington. Microsoft produces a variety of operating system and application software products. Its operating system products include MSDOS, Windows 3.1, Windows For Workgroups, and Windows 95, which currently dominates the market for PC operating system products. Its application products include many of the types of applications most commonly used by computer users.

With the growth of the Internet and the World Wide Web, consumer demand for Internet browser software products that provide access to these information sources has emerged and increased dramatically since 1994. Microsoft recognizes this demand and has developed successive versions of its Internet Explorer browser product to meet it. The initial version of Internet Explorer (version 1.0) was released in or around August 1995. Microsoft has since released three subsequent versions (2.0, 3.0, 4.0), in each case adding features and functionality to the browser product. Since at least December 1995, Microsoft has attributed great importance to capturing a major share of browsers and browser users. It has aggressively marketed and distributed its Internet Explorer product, not only by requiring OEMs, as part of and as a condition of receiving their Windows 95 license, to preinstall it, but also (as described below) by a variety of other means independent of the Windows operating system. Microsoft's aggressive and multifaceted marketing of the Internet Explorer browser reflects its intense competition with other, competing Internet browsers, primarily the "Navigator" browser produced by Netscape Communications Corporation ("Netscape").

Microsoft believes that the success of competing browsers poses a serious, incipient threat to its operating system monopoly. Indeed, as Microsoft fears, browsers have the potential to become both alternative "platforms" on which various software applications and programs can run and alternative "interfaces" that PC users can employ to obtain and work with such applications and programs. Significantly, competing browsers operate not only on Windows, but also on a variety of other operating systems. Microsoft fears that over time growing use and acceptance of competing browsers as alternative platforms and interfaces will reduce the significance of the particular underlying operating system on which they are running, thereby "commoditizing" the operating system. If this happens, PC OEMs' and end users' current, critical need for Windows, and thus Microsoft's monopoly power, would be reduced or eliminated and competition could return to the operating system market.

Prior Order of the Court

On July 15, 1994, the United States filed a civil antitrust Complaint to restrain Microsoft from using exclusionary and anticompetitive practices to market and distribute its PC operating system software, in violation of Sections 1 and 2 of the Sherman Act, 15 U.S.C. Among the practices alleged in the Complaint was that Microsoft had entered into license agreements with OEMs that required them to pay a royalty to Microsoft for each computer sold whether or not the OEM had preinstalled the Microsoft operating system product on the computer ("per processor" licenses). In addition, the license agreements often lasted three years or more, and OEMs were required to agree to large minimum commitments, with unused balances being credited to future license agreements. The effect of these practices was to exclude PC operating system competitors and monopolize the PC operating system market.

That action was settled upon consent and this Court entered a Final Judgment that enjoined Microsoft from, inter alia, imposing per processor licenses and binding minimum commitments on OEMs. To ensure that future Microsoft contracting practices with regard to OEMs did not replicate the anticompetitive effects of the challenged practices, the Final Judgment imposed additional prohibitions beyond banning existing practices. These prophylactic provisions included the following Section (hereinafter, "Section IV(E)(i)"):

Prohibited Conduct

Microsoft shall not enter into any License Agreement [with an OEM] in which the terms of that agreement are expressly or impliedly conditioned upon: (i) the licensing of any other Covered Product, Operating System Software product or other product (provided, however, that this provision in and of itself shall not be construed to prohibit Microsoft from developing integrated products). As the Competitive Impact Statement filed with the proposed Final Judgment made clear, this provision was designed to prevent Microsoft from attempting to extend or protect its operating system monopoly by conditioning its Windows license agreements on OEM's licensing or use of other Microsoft products.

Offense Charged

Microsoft has violated and continues to violate Section IV(E)(i) of the Final Judgment by requiring OEMs to license and preinstall Internet Explorer 3.0 as a condition of licensing Windows 95. Moreover, Microsoft will violate that Section in the future by requiring OEMs to license and distribute Internet Explorer 4.0 as a condition of licensing Windows 95.

The violation is clear: (a) Microsoft enters into "License Agreements" with OEMs that provide for the licensing of Windows 95, a "Covered Product" under the Final Judgment; (b) Microsoft's Windows 95 license agreements are currently conditioned on OEMs licensing Internet Explorer 3.0 and will be conditioned on OEMs licensing Internet Explorer 4.0; and (c) each version of Internet Explorer is an "other product" and not an "integrated product" within the meaning of the Final Judgment.

MICROSOFT DID NOT TIE IE AND WINDOWS

The June 21, 1999 direct testimony of Richard L. Schmalensee of the Sloan School of Management. In this excerpt, Schmalensee introduces the main points of his argument in defense of Microsoft's business practices. (The full text is available at www.microsoft.com/presspass/trial/schmal/schmal.htm.)

The inclusion of Web-browsing capabilities in Windows enhanced the value of Windows to consumers and ISVs and was therefore procompetitive. Microsoft's decision to integrate Web-browsing functionality into Windows was consistent with its previous business and technical decisions—over a span of 17 years—to integrate features into its operating systems, especially features that involve viewing, displaying, and locating files.

Plaintiffs, however, claim that operating systems and Web-browsing software are separate products sold in separate markets and that Microsoft illegally tied its Web-browsing "product" (IE) to its operating system product (Windows 95/98). In this section, I explain the economic fallacies in these assertions. In Part A, I show that there is not a separate product market for Web-browsing software. In Part B, I show that it would not have been efficient, and would not have benefited consumers, for Microsoft to provide a browser-less software platform. In Part C, I explain why Plaintiffs' contention that there is an incomplete overlap in demand for Web-browsing software does not demonstrate that there are separate product markets. In Part D, I explain why the fact that Microsoft provides IE-branded software for other platforms does not demonstrate that there is a separate product market for a browser-less version of Windows 95/98. In Part E, I explain why I believe that the approach taken by the D.C. Circuit Court of Appeals in the Windows 95 consent decree proceeding toward distinguishing integration from tying is based on sound economics and policymaking. Finally, in Part F, I show that Microsoft's integration of IE into Windows has none of the characteristics of ties that economists or the courts have found objectionable.

Before I turn to these economic arguments, it is worth pointing out that Plaintiffs' claim that Microsoft tied IE to Windows is wrong as a matter of fact. I understand that James Allchin will testify that the same components in Windows that provide Web-browsing functions also provide APIs for ISVs, parts of the user interface for non-Web functions of Windows, and access to information located off the Internet. Thus, "removing" IE from Windows is impossible. According to Netscape's outside counsel, in a letter to the DOJ,

> it is our understanding that it is simply not possible to delete any portion of [Internet Explorer], or of browser functionality, from Windows 98 as presently configured without severely interfering with the operating system.

Marc Andreessen, Netscape's other co-founder and one of the programmers of the original Mosaic browser, has testified in deposition (p. 235), that "[IE] is as inextricably integrated into the operating system as they wish to make it from a technical standpoint..." (p. 349).

No witness for Plaintiffs has testified that it is feasible to remove IE from Windows 98. Professor Edward Felten, one of Plaintiffs' technical experts, has testified that it is possible to write a software program that modifies Windows so as to prevent user access to the Web-browsing capabilities of Windows. (In other words, Professor Felten hides end user access to Web-browsing functionality but does not remove the platform capabilities of MS's Web-browsing software.) I fail to see the economic relevance of this claim, even if it were true. (I understand that Mr. Allchin will testify that Professor Felten has not fully disabled Web-browsing functionality and that his modifications cause serious degradation of Windows's performance.) The "fact" that it is possible to disable product features does not make the benefits of integration any less important to consumers, or negate the existence of integration. It is possible to disable air conditioning in new cars, too; that does not mean anybody would be better off if this were done routinely.

Therefore, based on the approach to tying adopted by the D.C. Circuit Court of Appeals—one with which I agree, as discussed below—there is no reason to conduct any further economic inquiry.

JUMP-STARTING THE DIGITAL ECONOMY (WITH DEPARTMENT OF MOTOR VEHICLES-ISSUED DIGITAL CERTIFICATES)

A publication of the The Democratic Leadership Council (DLC) & The Progressive Policy Institute (PPI), 600 Pennsylvania Ave., S.E., Suite 400, Washington, D.C. 20003, advocating a Department of Motor Vehicles licensing system for the use of digital authentication certificates for e-commerce.

Policy Briefing

June 1999 Marc Strassman and Robert D. Atkinson

The emerging digital economy promises high-productivity, low-unemployment, and increased standards of living. However, citizens, companies, or governments will be unable to fully realize these benefits until individuals can easily and securely authenticate themselves over the Internet.

Currently, few Americans can do this; that is, they are unable to fully represent themselves over the Internet in a way that securely tells other people and companies that they are who they claim to be and allows them to be taken seriously when they state their intentions. As a result, few companies or governments have developed applications that could use online authentication; and likewise, since few online applications require authentication, consumers have little reason to obtain the means to sign documents digitally. The Progressive Policy Institute (PPI) proposes that state governments should help jump start this process by providing digital certificates to all citizens who want them through state Department of Motor Vehicles (DMV) offices.

Just as we couldn't do business of any kind—educational, commercial, or interpersonal—if everyone walked around under a mask, it will be impossible to take full advantage of the Internet's power to collect, store, and distribute information, and therefore conduct various types of transactions, until each of us can authenticate ourselves online.

Authentication is an issue not unique to the Information Age. Medieval princes could secure and authenticate their documents with hot wax and a signet ring, ensuring that the message could not be tampered with without the recipient knowing it. Today, corporations and governments use official stamps and seals to signify the authenticity of the documents they issue. Similarly, digital signatures can be used to identify and authenticate documents and other files transmitted over the Internet.

The analogy between hot wax and signet rings and digital signatures is really very close. The engraved images on the signet rings were the product of some of that time's most advanced technology, engraving and metal work. Only the rich and powerful had access to the tools to insure the security and privacy of their data transmissions.

While digital signatures are based on an idea similar to the medieval signet rings, unlike the rings, digital signatures are potentially available to everyone. Using some of the latest computer and encryption technologies, digital signatures reduce a message to gibberish when it is tampered with, making it clear that the integrity of the document has been compromised, and allowing the recipient to disregard it.

Digital signature technology can be used to transfer into cyberspace the same, or a higher, level of assurance for legal and commercial purposes than has existed in common law, statutory law, and Uniform Commercial Codes for non-cyberspace transactions. By unambiguously and definitively establishing that a certain document has been "signed" by someone—or that someone has stated, indicated, and memorialized his or her intent to enter into an agreement of some type—digital signature technology makes it possible for binding transactions that cannot be repudiated to take place at a distance electronically. In short, digital signature technology enables today's e-commerce (online retailing) to flower into e-business and e-government (online transactions of a wide range).

What Are Digital Certificates and Digital Signatures?

To understand the applications and implications of digital certificates and digital signatures, it is important to understand what they do and how they do it.

First, think of the digital certificate as a pen used to write a digital signature. It is a unique digital code—a sequence of letters and numbers— that exists on a person's computer or smart card, that enables online identification. Certificates are provided by private companies that serve as certificate authorities (CA).

Then, think of a digital signature as the online equivalent to a signature you write with the pen. It is an encrypted and uniquely identified transmission that is attached to a signed document that becomes unintelligible if tampered with.

Here's How It Works

A person's digital certificate resides on their computer hard drive (or smart card). When a user wants to send a secure message or make any kind of online transaction requiring a digital signature, all he or she needs to do is access their certificate by clicking the appropriate icon on their Internet browser and entering their unique password. Employing the user's certificate, the computer will digitally "sign" a digest (an attachment to the document that the computer encrypts, or scrambles, using the sender's digital certificate). The signature is then added to the core document along with a "public key" that enables a certificate authority (CA), a trusted institution charged with supervising this process, to authenticate the signature.

When the message is received, the recipient checks with the CA to determine if the public key he or she has received is in fact the proper public key of the person sending the message. The recipient can then be assured that the message has indeed been "signed" with the claimed sender's digital signature. All of this, fortunately, is done by the computers in the background and is invisible to the user.

Using unique digital certificates to create digital signatures also allows both the sender and recipient to know for certain that the received message is identical to the sent message and that it hasn't been tampered with between its transmission and receipt.

It is important to note that the use of encryption for authentication does not raise the same law enforcement policy concerns presented by the use of encryption for confidentiality since only the digest, and not the message, is encrypted, and because the digest can be read by anyone using the sender's public key.

Online Authentication is Critical in Driving the Next Wave of E-Business and E-Government

Today, virtually all of the approximately $80 billion in annual consumer-based e-commerce involves transactions that do not require the user to authenticate him or herself. For example, buying a book from Amazon.com does not require that a person prove to Amazon that they are who they say they are; it simply requires that they provide a valid credit card number.

However, for a truly digital economy to fully emerge and provide the kinds of productivity and standard of living increases that are possible, a host of functions now conducted in-person or on paper must be able to migrate to cyberspace where transaction and processing costs will be a fraction of their current levels. For example, applying for a bank loan by phone costs $5.90, but using the Internet costs 14 cents. Similarly, the cost of a teller transaction at a bank is $1.07, while online it is one cent, and filing taxes online is at least 60% cheaper than filing paper copies.

A whole host of functions will depend on digital signatures if they are to be conducted online efficiently and on a widespread basis. These include applying for a loan or insurance; filing legal documents; applying for a permit, driver's license, passport, or other official government document; paying taxes; and even voting electronically. In short, a large share of transactions that now require our signatures for some form of identification could migrate to cyberspace—but only if digital certificates are in widespread use.

Yet, important as digital certificates and digital signatures are to the full development of e-business and e-government, they are not yet widely in use or even widely discussed. Melissa the MacroVirus got more publicity in three days recently than digital certificates have received in the last three years. The

n for this is that digital certificates and their relation to digital
is neither self-evident nor easy to understand. As a result, the
ua tend to shy away from the subject.

The complexity of these tools and the relative difficulty of obtaining them
has meant that few people have them. Without widespread adoption by
consumers, and with businesses apparently proceeding satisfactorily without
them, few companies or governments have developed applications that could
use online authentication. Likewise, since there are few online applications
that require authentication, consumers have little reason to obtain these
certificates. Moreover, putting digital certificates on smart cards (a credit
card-shaped piece of plastic that contains a microprocessor for performing
calculations, and a certain amount of computer memory for storing data)
only becomes a viable proposition if there are sufficient smart card readers in
use to attract enough users to support them. The chicken-and-egg metaphor
is the simplest way to describe the problem. The overall result is the one we
confront now: hardly any smart cards or digital certificates are in use
anywhere in the United States.

Nevertheless, increasingly powerful applications will become possible as
we move deeper into the Information Age, and many of them can only be put
in place, or put in place effectively, by using smart cards, digital certificates,
and digital signatures.

Accelerating the Adoption of Digital Signatures

As powerful and useful as digital signature technology is, there are certain
obstacles standing between where it is now and where it could be. Principally,
there is the problem of properly issuing the digital certificates upon which the
entire system depends. Candidates for digital certificates, like applicants for
driver's licenses, passports, or green cards, need at some point to present
themselves before trusted authorities and establish their identity, either on
the basis of a personal relationship with the trusted authority, or by presenting various types of documents that allow them to receive a digital certificate
in their own name.

Some say that the provision of digital certificates should be completely left
to the private sector. Clearly, the private sector needs to provide the technology, but it can also do this in partnership with government, the same way the
private sector helps the government accomplish many of its tasks, from
supporting a strong national defense to building roads.

Perhaps the most compelling reason why a government role is necessary
for a robust implementation of digital certificates relates to the very significant economic benefits derived from breaking out of the chicken-or-egg
conundrum faster than market forces alone are likely to be able to do. In
particular, the lack of knowledge of digital certificates—combined with the
cost and inconvenience involved in asking millions of citizens to present
themselves to separate digital certification agencies to establish their identity

and apply for a digital certificate—means that the use of digital certificates will develop only slowly, at best.

Not only will this mean that a host of e-business applications will be slow to develop, the same will also be true for many e-government applications. Perhaps the strongest motivation for states to make it easy for citizens to obtain digital certificates is that these will go a long way in enabling the electronic delivery of government services. If citizens could use their digital certificates to interact with state and local governments, the efficiencies resulting from online and electronic transactions would allow government to more than recoup the costs associated with providing the certificates. For example, citizens could apply for licenses and permits, file taxes, submit regulatory and other legal forms, and even vote online. Not only would state and local governments save millions, but citizen satisfaction with government would increase.

Fortunately, there already exists in every state and almost every community an agency whose job it is to establish and verify the identity of persons, and to capture that identity with a picture. This agency collects and stores what those in the identification business call "biometric indicators," such as height, weight, eye color, and hair color. They test your vision. They ask for your address. They make sure they know when you were born.

The Department of Motor Vehicles is already collecting quite enough information about each person to issue him or her a digital certificate. In fact, one can argue that it is the DMV that plays the baseline function of establishing authentication in the physical world. DMVs issue millions of driver's licenses and non-driver identification cards every year that people use to establish their identity in a myriad of applications. There is no reason why they shouldn't play this role in the cyber world. In fact, VeriSign, a leading provider of digital certificates, states: "Think of Digital IDs as the electronic equivalent of driver's licenses or passports that reside in your Internet browser and e-mail software." And indeed, the level of technological sophistication of the cards that embody these licenses varies from state to state. In many states, such as California, these cards include a magnetic strip, a digitized photo, and a surface hologram, designed to thwart illegal modification of the card or the data it holds.

Given that state DMVs already have sufficient data to issue digital certificates, that they already issue cards used for identification, and that they already employ sophisticated electronic and anti-tampering technologies, these agencies are well positioned to issue digital certificates as part of their ongoing citizen identification and certification functions. And since they already carry out their work on a rolling basis, with staggered renewals of their cards designed to balance the work flow, expanding their role to one of establishing identity in the cyber world would mean a gradual and smooth introduction of this technology.

To maximize the usability of such Government-Issued Digital Certificates (GIDCs), every citizen/customer/user who elects to could receive their driver's license on a smart card, which in addition to a photo and printed information on its surface, would also contain a microprocessor and have the capacity to accept and store a digital certificate. Citizens/users would select their own passwords and—from their own computer at home or at work, or from a publicly provided one in a school, library, or kiosk—generate and download their own unique digital certificate and private key.

This digital certificate would be a general-purpose digital certificate. There would also be room in the smart card for the user to allow other institutions, organizations, and companies to add "cardlets" that would entitle the cardholder to access his or her HMO records, to download e-cash, or to vote in elections. In order to assure security, these cardlets would be acquired by the holder on the basis of their general purpose digital certificate and whatever additional information other organizations or individuals required for access to specific databases or transaction opportunities.

People without computers could still use the digital certificates in their smart cards in various offline ways, such as for applying for government permits at a public computer kiosk. Credit card companies would perhaps become one of the organizations providing specialized cardlets for the smart cards. The potential of smart cards loaded with digital certificates to improve access, cut costs, and improve the efficiency of transactions that individuals conduct in the physical world is significant.[1]

In addition to providing the digital certificate to everyone on his or her driver's license or smart card, the state could also make the certificate containing the private key available directly to users to store on their computer(s) at home or at work, or both.

Likewise, this baseline authentication could be used to acquire other certificates that could be used for other purposes. Just as the driver's license is not the only means of personal identification, particularly for transactions with greater potential liability, other digital certificates issued by the private sector would also be used. With both smart cards and browser-based digital certificates, users would have private passwords that would prevent others from using their certificates to impersonate them in cyberspace.

As for the risk and liability questions surrounding the issuance and use of digital certificates in smart cards, there is a "defense in depth" approach that can effectively address this issue.

To start with, smart card and digital certificate users ("subscribers," in the industry jargon) are allowed to make up their own passwords. This reduces their need to write them down on their card. If they do make this mistake, and if their card is stolen and used fraudulently, the subscriber is liable, since the card issuer exercised due diligence in seeing that it would not be misused. However, since the leading digital certificate system employs a Certificate Revocation List (CRL) technology, once one of their subscribers reports his

---

[1]For example, one potential application for smart cards would be to enable consumers to register online for hotel reservations, and download the room key code to their smart card, which could then be used to enter the room without registering at the front desk.

or her card lost or missing, it can be revoked immediately, and anyone trying to use it will not be able to do so. This is like revoking a credit card, only faster and more certain.

The ability to instantly revoke a certificate also comes into play in the case of cards that are stolen and then attacked to discover their password. In addition to the revocation protection, the cards themselves are resistant to forced intrusion. Ten thousand computers working simultaneously for 22 hours are required to break a 56-bit key. Current cards employ 128-bit keys, and future versions will feature 256-bit keys, so it will take much longer to intrude into these—far longer than the time it takes to revoke the card entirely.

As for the previously mentioned private-sector participation, it makes sense for each DMV to outsource the actual provision of the digital certificates and the smart cards, as well as the management of the certificates, to one or more private companies with established track records in developing, deploying, and managing digital signature technology. In the same way that state governments hire private companies to supply copying or phone services, or even today's driver's licenses, they would contract with established digital signature technology companies to provide the necessary components required to introduce and maintain the processes that constitute the digital signature system. Moreover, they could choose whatever parameters and technologies for authentication they think work best and are most cost-effective. In fact, different states may use different technologies.

Finally, the fact that DMVs would issue these cards would in no way prevent individuals who would rather obtain certificates from private providers from doing so. Rather, it would simply make it easier for individuals to obtain them. In addition, just as individuals now use multiple forms of identification (such as passports, birth certificates, and witnesses) for certain transactions—especially more sensitive ones (e.g., papers that need to be notarized)—some individuals would likely obtain multiple digital certificates that could be used in combination or individually, but the DMV-issued certificate serving as a baseline.

A Threat to Privacy?

Aren't digital certificates a step toward a national ID or a potential threat to privacy? Personal privacy has long been a core American value, and the proliferation of modern database technology has done nothing to eliminate this concern. In fact, it has only made it a more pressing matter.[2] Banks, merchants, HMOs, and the government all possess a lot of data about us and our habits, a fact that will not change in the presence or absence of a satisfactory means of issuing digital certificates.

---

[2]Court, Randolph H., and Robert D. Atkinson. March 1999. Online Privacy Standards: The Case for a Limited Federal Role in a Self-Regulatory Regime, Washington, D.C.: Progressive Policy Institute.

Moreover, obtaining digital certificates from the DMV would be voluntary, and the state government would not itself serve as the certificate authority or know the passwords individuals choose to access the certificates. Also, just as driver's licenses are issued by states and not the federal government, under this proposal states would also issue digital certificates.

Finally, just as there are some transactions in the physical world that are anonymous and some that require identification, the same is true in the cyber world. Through the process of "anonymous authentication"—developed to allow voters to be authenticated online while maintaining the confidentiality of their electronic ballots and preventing their choices from being personally associated with them—other subscribers can also authenticate themselves as necessary while preserving certain aspects of anonymity in various other types of transactions. It will be important for state and local government to not require personal identification online when simple authentication will do. For example, a county may require that someone prove they are a resident before accessing a data base. In this case, a digital certificate would certify only that the person is a resident without revealing his or her identify. Fortunately, the technology is flexible enough to easily accomplish this. In addition, DMVs and the private digital certificate providers should establish a code of privacy that keeps the data they collect private. Overall, clearly thought out and reasoned government policies should prove sufficient in most cases to address these and other similar concerns.

Summary

It would not be an abrupt change for state DMVs to begin issuing driver's licenses on smart cards, and to provide the means for each citizen who wants to create and store a digital certificate on that card. It would be, instead, an incremental modernization which will set the stage for a rapid advance in efficiency and cost-saving within state government, for an explosion of e-commerce, and for the facilitation of countless everyday tasks for every certificate holder.

References

Berners-Lee, Tim. 1996. "The World Wide Web—Past, Present, and Future." Distinguished Fellowship Award Speech of the British Computer Society. http://www.bcs.org.uk/news/timbl.htm. (3 February 2000).

Department Leadership Council and the Progressive Policy Institute. 1999. "Jump-Starting the Digital Economy (with Department of Motor Vehicles-Issued Digital Certificates)," June 1999. http://www.dlcppi.org/texts/tech/jumpstart.htm. (3 February 2000).

National Telecommunications and Information Administration. (U.S. Department of Commerce). 1998. "Memorandum of Understanding between the U.S. Department of Commerce and Internet Corporation for Assigned Names and Numbers." http://www.ntia.doc.gov/ntiahome/domainname/icann-memorandum.htm. (3 February 2000).

Schmalensee, Richard L. 1999. Testimony in Civil Action Numbers 98-1232 and 98-1233, 11 January 1999. http://www.microsoft.com/presspass/trial/schmal/schmal.htm. (3 February 2000).

United States District Court for the District of Columbia. 1998. "Petition in the United States District Court for the District of Columbia, January 21, 1998." http://www.dcd.uscourts.gov/microsoft-findings.html. (3 February 2000). (Also available at http://www.usdoj.gov/atr/cases/ms_index.htm along with supporting and subsequent documents.)

CHAPTER SEVEN
Career Information

From online researchers and interactive television writers to librarians and inventory control clerks, the outlook for employment in the information technology (IT) sector will continue to expand for years to come. As we are coming to realize, almost everyone who will work in the developing information-based economy will be using the digital tools now being created by computer scientists and engineers. Everyone, from user to developer, will add to the utility and efficacy of the next generation of computing environments, but the jobs profiled in this chapter concentrate on those that can be most readily grouped into the computer science category.

In 1996, the Bureau of Labor Statistics (BLS) predicted which occupations would have the greatest rate of increase in the United States between 1996 and 2006. The top two job categories were both computer science related. In 1996, there were 215,700 computer engineers, by 2006 that segment of the work force should number around 451,000. This 109% increase is slightly ahead of the jump of 103% that will occur in the second place category: systems analysts and electronic data processors. The 1996 figure of 505,500 will likely increase to 1,025,100 in 2006.

The February 1, 1999, issue of *Newsweek* had a special section on careers for the turn of the century. That report reiterated the BLS findings. Six of the ten "hot" jobs listed in that article had to do with computer science. They included chief information officer with a predicted salary of $100,000–$200,000, software development manager ($60,000–$80,000), computer systems architect ($60,000–$100,000), database administrator ($60,000–$80,000), director of e-commerce ($50,000–$80,000), and Webmaster ($50,000–$70,000). The employment outlook for people who want to help create and support the digital age is bright indeed.

In 1998, the National Software Alliance estimated that about 137,000 software workers are needed to fill openings on an annual basis. However, the trend in training for these positions is not a positive one. The number of computer science degrees awarded between 1988 and 1995 *decreased* by 42%. The result is that only one-quarter of the jobs for new or replacement positions are being filled by qualified workers. This fact has resulted in a crisis for companies under pressure to develop new and improved products at the heady pace people have come to expect from the industry.

To meet this demand, especially for programmers and systems managers, companies are looking to graduates in related fields of study. As one might expect, many employers are happy to take math, physics, or electrical engineering majors into entry-level positions so that they can train them to perform tasks specific for that company. What may be a surprise to some readers is that recruiters have their sights set on motivated "career changers" too. They may be just graduating or looking to take a new career path after years of successful experience working in an area that one wouldn't necessarily associate with computer engineering, programming, or any directly applicable technical skills per se. Driven by the need to fill thousands of positions, these employers are anxious to identify motivated individuals who have a proven track record in employment or study, but who also have "played around" with or demonstrated an affinity for PCs at previous jobs, school, or at home. By all indications, the outlook for those considering entering the IT fields will remain very positive for the next decade.

CATEGORIES OF INFORMATION TECHNOLOGY JOBS

Since the industry discussed here is so new, in some instances jobs and job categories are still being defined. What follows are the basic designations and descriptions for the most common task sets found in this particular segment of the economy. After naming and defining the position, the standard schooling, training, and experience path is included. Please note that this list is not in hierarchical order, rather in alphabetic order. It also does not include the many possible positions that are evolving from the expanding capabilities of digital telecommunications systems. Broadband systems installers, online librarians, and search engine researchers, are just a few of the obvious possibilities as we move into the new century and a more mature phase of the IT revolution.

Chief Information Officer

The Chief Information Officer (CIO) is a senior executive of a large enterprise responsible for all aspects of a company's information technology systems. This person is usually the one with the best insight into how to align the business goals with the proper technological solution for any given situation. Since the CIO is in a high management position, reporting directly to the Chief Executive Officer or the Chief Operating Officer, the most important duties of the office are strategic, having to do with implementing and managing information systems that oftentimes redefine the methods and manner in which a company will conduct all aspects of its business. As information technology becomes more important to the business sector, the influence of the CIO increases.

While this visioning function is most important, it is also true that the CIO may oversee the day-to-day operations of the information technology component. This includes responsibility for purchases that include everything from the hardware and software required to move and use information within and without the enterprise, to the contracting of ancillary digital services that support the company's mission. They are oftentimes in the best position to direct Internet and Web initiatives.

Basic skills and experience required to be a chief information officer will differ depending on the industry and company involved. In all cases, however, it is desirable to have a strong business orientation. This can be achieved with a combination of consulting or employment experience in the industry under consideration. A degree in business administration along with computer coursework is very desirable. Technical skills in managing IT solutions and successful implementation of new information processes in a related organization will also be scrutinized by a potential employer.

Computer (Electrical) Engineer

Tasked with designing and testing new semiconductors, circuit boards, whole computer systems, and peripheral devices, the computer engineer seeks to create computing devices that operate in an increasingly efficient manner. Computer engineers can be in charge of development in research labs, design specifications for new machines (device development, circuit design, hardware engineering), run production lines, and manage the output for factories, or function as quality control experts who put their stamp of approval on finished products.

Computer engineers often go through a well-defined process in the development of new hardware. They first make a determination regarding the exact function they want their new machine to perform. They design components (often using computer-aided design [CAD] applications), build a prototype, and test the functionality of their idea. They then determine the economy, efficiency, reliability, safety, and effectiveness of their products. In this process, the engineer is aided by knowing programming languages like Fortran, C, or C++, as they work on projects from mainframes and supercomputers to robots and artificial intelligence systems.

The basic requirements for employment as a computer engineer include a bachelor of science (BS) in electrical or computer engineering. Additional graduate courses, specialized training, or a masters degree are also advantageous to keeping abreast of advances in this field.

Computer Support Careers

Increased reliance on computers and digital systems in every aspect of society has given rise to another important segment of the working population, the computer support experts. Generally their work falls into two distinct categories. The first group is charged with testing, marketing, or maintaining computers. This they do as independent consultants or as part of a company. The second group is charged with end user support. These are data processing and information systems professionals who make certain that clients and users are successful with particular products.

Examples of support personnel in the first designation include sales and marketing positions like product managers, marketing managers, public relations representatives, and sales associates. Customer services people take over once the sale is made. These positions include corporate product support personnel who work with large organizations when they adopt or upgrade systems. Other specialist positions, like quality assurance control technicians and independent consultants are also included in this group.

In the second case, support personnel like IT analysts, PC analysts, management of information system (MIS) directors, and network operations managers are part of the larger management of information systems group. In the first two cases, consultants monitor and report on changes and best practices in the IT industry. MIS directors and network operation personnel are called in when a new hire needs training or a veteran needs help troubleshooting a problem on applications or hardware within the corporate structure. Help desk personnel and technical support people are also on staff to answer questions and to give direct user support to individual clients who have purchased a product.

These computer support positions require very different levels of training and skills to achieve employment. For example, the MIS Director position will typically require a certification or degree with at least an MIS minor. Coursework in either of these programs will include business basics as well as project development, basic programming, networking, and information systems classes. However, while some college work is usually preferable for employment as a technology support or help desk person, essential training will come on the software or hardware that is being supported. Often, companies have their own programs for this education.

Computer System Architect

The person who fills this position in a large enterprise is sometimes referred to as the "jack of all trades" of the information technology industry. The system architect fills many roles in the organization like guaranteeing adaptability across different computer systems and understanding technical possibilities and limitations. Most importantly, this manager is responsible for providing direction to the company leadership in terms of how the business goals can be best facilitated with existing and future digital technologies. Because of the broad spectrum of areas in which oversight and advice must be provided, the system architect is required to have a working knowledge of many computer science areas. These include

1. Network infrastructures and the hardware (routers, hubs, switches, etc.), software protocols (HTTP, FTP, SMTP, etc.), and administration techniques (firewalls, proxies, etc.) that define the modern distributed computing enterprise.
2. Server technologies that include the emerging microprocessors and their scalability and clustering capabilities, operating systems, memory, file, and print services.
3. Security services that monitor and assure privacy and secure transactions software development in all of the current languages (C, C++, COBOL, Java, Pascal, PowerBuilder, Visual Basic, Forte, Dynasty, and others).

Database Administrator

The database administrator (DBA), or manager, is responsible for the design and management of databases and for the evaluation and implementation of the deployed system. Modern databases like those produced by ORACLE and Microsoft's SQL Server are the engines behind many Internet and intranet (internal web network) sites. Working in conjunction with a large organization's system analyst and architect, the database administrator typically optimizes access to data for both employees and clients. This task can be challenging when one considers the large demand on computing resources that occur when thousands of users attempt to access a site at the same moment. The DBA must be able to predict how and when data will be called for. By taking these, and other factors like hardware and connectivity speeds into account, a system is less likely to fail when placed under extreme loads.

The robust nature of database system development is becoming more critical as the amount of commerce taking place over the Internet increases. A specialized version of the database administrator, the e-commerce manager, is evolving from the ranks to take responsibility for this change in the way people do their shopping. Usually, the person in this position will have substantial experience in database design and administration along with a good working knowledge of HTML coding, JAVA and the various general and specific requirements for selling their company's products via Internet Explorer and Netscape Web browsers.

In general, the DBA must have successful experience, and in most cases, certification, in either Oracle or SQL Server databases. Other database experience and knowledge is helpful as well.

Data Security Specialist

These are the people who know how to police large computer systems so that the data they hold is protected against tampering or loss from any hostile or malicious external force. That force might be from such things as storms, fire, power loss, or flooding. It can also involve computer hackers, internal thieves, viruses, or fraud. In order to be effective in this work, security specialists need a background in programming, an understanding of systems, and familiarity with telecommunications and networking protocols of all types. They must also understand the nature of the business enterprise in which the computer system is deployed.

This highly specialized work requires individuals with a great deal of experience in computer science and engineering, and with an emphasis on other technical skills specific to their particular business. Security experts will almost always be on the staff of large corporate IT departments, but it is not uncommon to find consultants working in this field, especially in relationships with small businesses and virus outbreaks.

Electronic Data Processing Auditor

While they are also concerned with the security issues that could affect an enterprise's data structures, the main role of the electronic data processing (EDP) auditors is that of data accountant. These are the people who make certain that the company's information systems operate as they were designed by the systems analysts to operate. They oversee all aspects of the operation that influence how data and information is stored and accessed. This involves checking data structures via the programming and ascertaining that assurance and security plans are adequate.

The following minimum requirements are from an actual advertisement for an open EDP auditor position, posted by a corporation wherein this professional would most typically work:

> Graduation from an accredited college or university with a bachelor's degree in accounting, banking, auditing, economics, finance, statistics, mathematics, business administration, public administration, or computer science; AND three years of professional accounting, auditing, financial examination or systems analyst experience which included at least one year of EDP auditing experience.

Interface Designer

The interface between the human and the machine is a new area of research, but it is one that is seeing a new emphasis as more powerful microprocessors and emerging display and input technologies become more cost-effective. The person who strives to improve the ease with which humans communicate with their intelligent devices is called the interface designer. This person's work can focus on ergonomic issues like the way the wrist or hands are used for input, the effect of different displays on visual response, or the way portable or wearable devices fit into a person's daily routine. It can also focus on the way large amounts of data are tracked, stored, and retrieved, so that its use is made efficient and comfortable for the user. In almost every case, the interface designer is working in a university or corporate research facility like Xerox PARC or Bell Labs.

Some typical steps an interface designer might go through include the following:

- A study is made about the way machines present information in conjunction with how humans use information and how they might want to use it in the future.
- A conceptual design (a storyboard) of a possible interface that could be prototyped is constructed.
- A working prototype is created so that it can be used to evaluate whether the hypothetical design is on the right path.
- If the designer is not an artist or a writer, an artist or writer may be consulted to develop a look and feel for the eventual machine.
- Testing further and gathering user feedback, critical to the eventual deployment of an improved interface, is completed.

A BS or BA in computer science or in information studies would prove very helpful in this position. In addition, experience in psychology, kinetics, graphic arts, and communications theory is desirable.

Multimedia Developer

The increase dependence and spread of the Web as a new medium for the delivery of information has meant a new demand for interactive multidimensional presentations that catch the users eyes, ears, and interest. Multimedia developers are still inventing the tools of their trade and defining the nature of their output. They are, indeed, the artists of the digital age. Their work, which can incorporate video, text, audio, and interactive and animated graphics in creative displays is most often witnessed on the Web, but can be distributed via CD-ROM, DVD, tape, etc. As the technology moves on, so do the way these people use it.

Multimedia developers are typically self-taught. They may come to the discipline from fields as diverse as education, television, writing, computer science, or the graphic arts. There are degree programs offered now in the newest media fields, and training availability is likely to grow in the future. As content becomes more important to the new Internet-based media outlets, more and more multimedia artists will find employment with corporate entities. But the majority of this work is presently taking place in the homes of freelance workers who find the freedom of "independent contractor" status to be conducive to the production of imaginative products.

Programmer and Software Engineer

The code that becomes computer applications is written by programmers. In a large operation, the programmer is handed an assignment based on specifications developed by the systems analyst. Programmers can be found working on applications or developing new systems. There are also systems programmers who create new operating system software or maintain and improve established operating system software. Companies like Microsoft and Oracle employ thousands of programmers, and they are always looking for additional people to fill their cubicles. But there is also a vast army of freelance programmers, writing code in their homes or in university computer labs, in an effort to create the next Netscape or to simply improve one of the thousands of open source programs currently in distribution.

In most cases, the process that programmers go through in their work is organized in the following manner:

- Coding is the nuts and bolts of typing in specific program instructions, according to the language designated for use in the design process. The work at this point must be done in a manner that introduces as few mistakes (a.k.a. "bugs") as possible.
- Compiling is the process by which the original language of the program (high-level) is converted into the native language of digital machines, i.e., binary code. This is the first indicator of a potential program's viability. Should the code be flawed, it will fail to compile and the compiler utility will generate a bug report.
- "Debugging" using the bug, or error, report as a roadmap, the programmer will then attempt to fix the offending faulty code and recompile.
- Testing the application by running real data in everyday situations is a crucial step. Further bugs that crop up must then be dealt with before the application can be deemed stable.
- Maintenance is an ongoing and thorough process in the better software labs. Once a program is released to the general public, it almost always acts "buggy" on certain machines and given specific configurations of other resident software. Continual reworking of the code often calls for the release of updated versions of software.

A related job is that of the software development manager, who must meet all the skill requirements for programming that a software engineer has, but in addition, must have leadership abilities that allow for direction of teams of programmers working toward an assigned goal. The manager's knowledge of operating systems, programming languages, and techniques to build successful applications comes together with interpersonal management skills, creating the lead person on the development team. Assigning duties to program team members, providing a clear vision of the end product, monitoring progress, testing results, and interfacing with the company's marketing department and end users to assure compatibility and functionality are the key responsibilities of the development manager. A project can take anywhere from a few weeks to a year or more to complete.

A bachelor of arts or science in computer science or in information systems is required for many programmer positions. But as was alluded to previously, often programmers with an associate of art (AA) degree or appropriate experience level will have little trouble finding employment.

Systems Analyst

Together with the software engineers, systems analysts manage the development processes involved in creating new hardware and software. They are usually employed in a corporate research and development setting, but many independent systems analysts contract for particular projects as freelancers. The system analyst, especially, will oversee both hardware and software research, while the software engineer tends to stay in the programming realm. Typically, the long process that results in the release of a new hardware or software product includes these steps:

- Problem is defined. Goals and objectives are set as a result of interviews and research that other designers and users have done.
- Program (or device in the case of hardware) is designed. This process involves a time-consuming identification of all of the crucial elements that must be integrated to achieve a solution for the identified problem.
- Flow chart diagram is created. This and other analytical tools attempt to paint a clear picture of data flow through the new product, with an eye on economic viability and sustainability.
- Specific files and elements are identified including the user interface, network functionality, security, etc.
- After completion, the program or device is tested and evaluated.

A BS in computer science or a BS or bachelor of art (BA) in a related field of inquiry, plus some prior practical experience, are desirable attributes for a systems analyst. Specific computer platform knowledge and a comfort level with programming are extras that recruiters will look for. This is expected to be one of the fastest growing fields in the next five to ten years.

Virtual Reality Researcher

These professionals create computer generated worlds (virtual environments) that can be entered by users who wear special gloves, head gear, and other sensory devices. The idea is to be able to immerse the user in data that is displayed in new and interesting ways.

Virtual worlds are being used to display impossible-to-observe aspects of the human anatomy, build 3-dimensional models of planet surfaces, control robots, build airliners, and design homes. Virtual reality (VR) researchers are creating machines, hardware, and applications that have achieved these feats and many others. The future for people in this job classification continues to be very bright, whether employment comes from a large research/manufacturing facility like Boeing, where their 777 jet was designed using VR systems, or in a university laboratory.

Because VR is such a complex endeavor, people who choose this field would be wise to have a background in learning theory and knowledge development, psychology, education, computer science, and electrical engineering.

Webmaster

One of the new jobs that have arisen with the success of the World Wide Web (Web), the Webmaster manages the Web server of a particular site on the Internet. This can mean making certain that the server software and hardware is operating up to its maximum efficiency, maintaining security and viability of data, creating Web-based content, and tuning other applications so that they can serve up data through the server. This work is now done in a home office, at the office of an Internet Service Provider (ISP), or in a corporate office where the company has established an online presence.

It is important for the Webmaster to be constantly upgrading skills and keeping current regarding the new applications and tools for maintaining sites, authoring content, and providing a media rich experience for site visitors.

While programming skills (especially in Java and Perl) can be a desirable attribute for Webmasters to possess, an interest in computers, graphics, design, multimedia, and the Internet may be even more important at this early stage. Most Webmasters are self-taught, gaining experiences by running Web sites. In the future, the ability to create and maintain robust e-commerce applications appears to be of critical importance.

TRAINING AND EDUCATION OPTIONS

An IT analyst recently said this about the job market in the United States:

> It's a tight labor market in most of the U.S. for computer programmers and other MIS [managed information systems] professionals—no doubt about it. Companies are going crazy trying almost anything to find even semiqualified applicants to fill the growing

number of jobs in this field. They are outbidding each other on salary, offering signing bonuses as if programmers were major-league athletes, and providing benefits and perks ranging from cappuccino machines and concierge services to on-site day care and neck massages at your desk (Gordon 1997).

There are even reports circulating about technically advanced high school students being offered incentives to join large software firms with starting salaries near the $60,000 range. The message is, almost anyone can look at the possibility of employment some place in this sector. Career changers might look to supplemental course work at local colleges, online distance education offerings, and some of the prepackaged certification programs to enhance their chances of being considered for the IT track they really want. Or, they might consider going with a company that values eclectic experiences and so-called soft or people skills where the policy is to train new hires in fast track, quick start, training programs. Then too, one might want to consider the time-honored route of obtaining an accredited college degree in computer science, information technology, electrical engineering, or mathematics.

While it is true that computer science (CS) departments are some of the newest additions to university faculty hierarchies, more and more good degree programs are being established around the United States and Canada. Massachusetts Institute of Technology (MIT), McGill, the University of Illinois, Stanford, University of California—San Diego, and University of Southern California are a few of the well established and most honored CS departments in the world. But, in most cases, it is easy to find local programs offering the core educational elements for obtaining a degree in a computer related field. A course of study for the first two years in a typical college includes

- Programming
- Operating Systems
- Software Engineering
- Data Structures
- Hardware, Logic, and Machine Languages
- Numerical Methods

The final years of study are dedicated to working through a choice of electives that would generally include

- Advanced Software Engineering
- Operating Systems
- Computer Architecture
- Computer Graphics

- Databases
- Management Information Systems

It may also prove beneficial to understand the requirements for the same degree at one of the premier computer research institutions. Carnegie Mellon University in Pittsburgh has one of the oldest CS departments, started in 1965. Today, it is considered one of the best places to study computing in the world. The following, from Carnegie Mellon's own 1999 catalogue materials, describes only the computer-specific coursework required for their bachelor of computer science degree:

- Introduction to Programming and Computer Science
- Fundamental Structures of Computer Science I
- Fundamental Structures of Computer Science II
- System Level Programming
- Great Theoretical Ideas in Computer Science (or Discrete Mathematics)
- Algorithms

One of these applications courses:

- Artificial Intelligence: Representation and Problem Solving
- Robotic Manipulation
- Computer Vision
- Software Engineering
- Database Applications
- Computer Graphics

One of these fundamentals of programming courses:

- Programming Languages Design and Processing
- Formal Languages and Automata
- Models of Software Systems
- Basic Logic
- Computability and Incompleteness

One of these systems programming courses:

- Operating Systems
- Computer Networks
- Concurrency and Real-Time Systems

Two computer science electives must be chosen, in addition to all of these mathematics/statistics courses:

- Differential Calculus
- Integral Calculus

- Integration and Differential Equations
- Calculus of Approximation
- Concepts of Modern Mathematics

One of these statistics tracts is also required:

- Probability Theory and Random Processes
- Introduction to Probability and Statistics I & II
- Probability and Mathematical Statistics I & II

References

Bureau of Labor Statistics. 1998. *Occupational Outlook Handbook.* http://www.bls.gov/ocohome.htm. (10 October 1999).

Gordon, Gil. 1997. "How to Fill Jobs, Part I: The Stomach as Career Director." *Telecommuting Review: the Gordon Report*, 1 August 1997. Monmouth Junction, NJ: Gil Gordon Associates.

Microsoft IT Career Resource Center. 1999. http://www.microsoft.com/train_cert/itcr/default.htm. (21 October 1999).

CHAPTER EIGHT
Statistics and Data

T he graphs and tables in this chapter provide snapshots of the state of various computer-related issues addressed in previous chapters. They show a variety of aspects ranging from purely technical factors to the social issues involved in the deployment of digital technology and communications systems.

Table 8.1 is a charting of the advances that have occurred in the Intel microprocessors (from 1979 to 1999, which fulfill the general predictions of Moore's Law as discussed in chapter 1. Various parameters, including clock speed and bus size are represented.

Table 8.2 shows the growth in the number of Internet domains (computers with individual Internet protocol (IP) addresses connected to the network) between the years 1981 and 1999. Figure 8.1 is a graph of the same data from the years 1991 to 1999, provided for a dramatic visual representation of the remarkable growth of the Internet over this time frame.

Figures 8.2, 8.3, and 8.4 relate to the development of a broadly distributed computational network as discussed in chapter 2. Figure 8.2 is a graph that shows the demand for computer time by scientists accessing the supercomputing facilities around the United States. This exponential growth in demand for computational resources is the basis for the deployment of the "Grid" and other "Internet2" architectures. Figures

8.3 and 8.4 are maps representing two of the emerging information infrastructures.

Tables 8.3, 8.4, 8.5, and 8.6 are included for the data they provide on the trends in higher education from 1995 to 1997. Table 8.3 is a display of the number of postgraduate degrees awarded in selected science and engineering disciplines. Included are pertinent data about computer science graduates. Figure 8.4 is a charting of the masters and doctorates granted in related computer areas over that same time period. Figure 8.5 looks at gender differences in selected science degrees awarded in 1997. Figure 8.6 is an overview of all science (engineering and health fields included) graduate students enrolled in institutions of higher education in the United States between 1990-1997.

Where are the jobs going to be once students graduate? Table 8.7 is a chart showing the Bureau of Labor Statistics projections for the growth of jobs in specific sectors of the American economy. This particular data include only the top categories in growth expected by the year 2006. Note that computer-related areas are heavily represented.

In Table 8.8, provided with permission from its author, Gunther Ahrendt of GAPCON Corp., Ahrendt lists the most powerful computing sites in the world. Reproduced here is the top portion of the list that was published January 7, 2000.

MICROPROCESSORS

Table 8.1. Evolution of the Intel Microprocessor Specifications 1979–1999

Category Designation	Family Generation	Date Introduced	Data/ Address Bus Width	Internal Cache Size[1]	Memory Bus Speed[2] (MHz)	Internal Clock Speed[3] (MHz)
8088	First	1979	8/20 bit	None	4.77-8	4.77-8
8086	First	1978	16/20 bit	None	4.77-8	4.77-8
80286	Second	1982	16/24 bit	None	6-20	6-20
80386DX	Third	1985	32/32 bit	None	16-33	16-33
80386SX	Third	1988	16/32 bit	8K	16-33	16-33
80486DX	Fourth	1989	32/32 bit	8K	25-50	25-50
80486SX	Fourth	1989	32/32 bit	8K	25-50	25-50
80486DX2	Fourth	1992	32/32 bit	8K	25-40	50-80
80486DX4	Fourth	1994	32/32 bit	8K+8K	25-40	75-120
Pentium	Fifth	1993	64/32 bit	8K+8K	60-66	60-200
MMX	Fifth	1997	64/32 bit	16K+16K	66	166-233
Pentium Pro	Sixth	1995	64/36 bit	8K+8K	66	150-200
Pentium II	Sixth	1997	64/36 bit	16K+16K	66	233-450
Pentium III	Sixth	1999	64/36 bit	16K+16K	100	500-550

[1]Internal cache refers to the fast memory module that holds recently accessed data.

[2]Memory bus speed is the rate by which data move in megahertz (MHz) through the conductors connecting the central processing unit (CPU) to its components.

[3]Internal clock speed is the fundamental rate in cycles per second at which a computer performs its most basic operations.

Source: Intel Corporation, www.techweb.com. (10 October 1999).

INTERNET DOMAINS

Table 8.2. Number of Internet Domains from 1981–1999 [*]		
Date	Hosts	Source
Aug-81	213	host table
May-82	235	
Aug-83	562	
Oct-84	1,024	
Oct-85	1,961	
Feb-86	2,308	
Nov-86	5,089	
Dec-87	28,174	old domain survey
Jul-88	33,000	
Oct-88	56,000	
Jan-89	80,000	
Jul-89	130,000	
Oct-89	159,000	
Oct-90	313,000	
Jan-91	376,000	
Jul-91	535,000	
Oct-91	617,000	
Jan-92	727,000	
Apr-92	890,000	
Jul-92	992,000	
Oct-92	1,136,000	
Jan-93	1,313,000	
Apr-93	1,486,000	
Jul-93	1,776,000	
Oct-93	2,056,000	
Jan-94	2,217,000	
Jul-94	3,212,000	
Oct-94	3,864,000	
		adjusted counts
Jan-95	4,852,000	5,846,000
Jul-95	6,642,000	8,200,000
Jan-96	9,472,000	14,352,000
Jul-96	12,881,000	16,729,000
Jan-97	16,146,000	21,819,000
Jul-97	19,540,000	26,053,000
Jan-98	29,670,000	new domain survey
Jul-98	36,739,000	
Jan-99	43,230,000	

[*]Domains connected to the Internet have been surveyed since 1987. As the Internet changed, the methodology employed had to change also. That adaptation is reflected in the third column. Documentation on the methodology is available at the Network Wizards Web site.

Source: Internet Software Consortium. http://www.isc.org/. (8 June 1999).

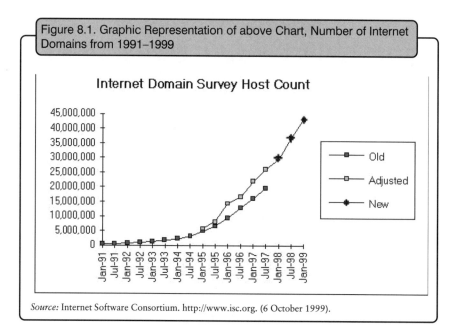

Figure 8.1. Graphic Representation of above Chart, Number of Internet Domains from 1991–1999

Source: Internet Software Consortium. http://www.isc.org. (6 October 1999).

THE NEXT GENERATION INTERNET

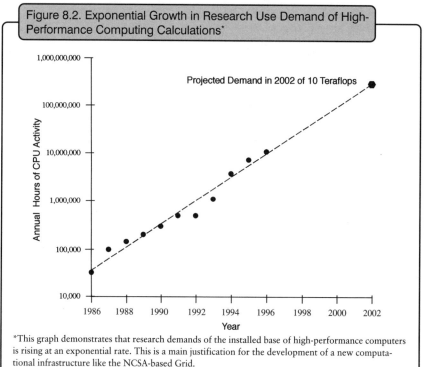

Figure 8.2. Exponential Growth in Research Use Demand of High-Performance Computing Calculations*

*This graph demonstrates that research demands of the installed base of high-performance computers is rising at an exponential rate. This is a main justification for the development of a new computational infrastructure like the NCSA-based Grid.

Source: Graph created from data available from the NCSA and Quantum Research database.

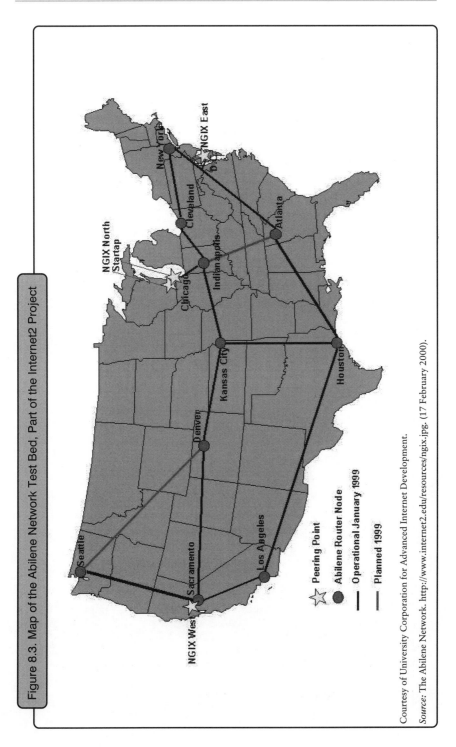

Figure 8.3. Map of the Abilene Network Test Bed, Part of the Internet2 Project

Courtesy of University Corporation for Advanced Internet Development.

Source: The Abilene Network. http://www.internet2.edu//resources/ngix.jpg. (17 February 2000).

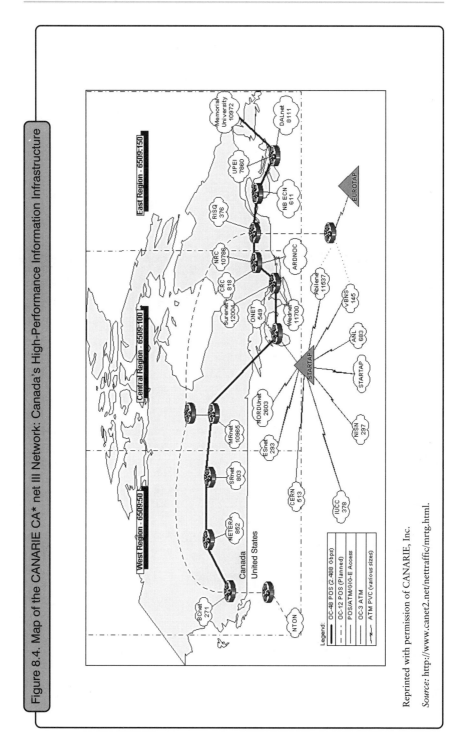

Figure 8.4. Map of the CANARIE CA* net III Network: Canada's High-Performance Information Infrastructure

Reprinted with permission of CANARIE, Inc.

Source: http://www.canet2.net/nettraffic/mrtg.html.

TRENDS IN HIGHER EDUCATION

Table 8.3. Total Science and Engineering Postdoctorates Awarded in the United States by Discipline, 1995–1997*

Total S&E Postdoctorates

Academic Discipline	1995	1996	1997
Total of All Academic Disciplines	35,926	37,107	38,043
S&E Total (including medical/other life sciences)	35,926	37,107	38,043
S&E Total (excluding medical/other life sciences)	26,183	26,591	26,917
Engineering	2,641	2,674	2,945
Physical Sciences	5,863	5,839	5,907
Geosciences	845	861	938
Math and Computer Sciences	475	576	623
Computer Science (alone)	213	250	318
Life Sciences	25,144	26,119	26,707
Psychology	582	594	567
Social Sciences	376	444	356

*The data in this chart were gathered by the National Science Foundation and include the latest research on postdoctoral science and engineering (S&E) degrees.

Source: NSF WebCASPAR Database System. http://caspar.nsf.gov/.

Table 8.4. Total Master's and Doctorate Degrees Granted in Selected Engineering and Science Disciplines 1995–1997*

Graduate Student Survey —Academic Discipline	Highest Degree Granted	1995	1996	1997
Sciences and Engineering (excluding health fields)	Total of All U.S. Institutions	422,555	415,363	407,644
Sciences and Engineering (excluding health fields)	Doctorate	372,994	366,805	359,476
Sciences and Engineering (excluding health fields)	Master's	49,561	48,558	48,168
Sciences (excluding health fields)	Total of All U.S. Institutions	315,356	312,140	306,636
Sciences (excluding health fields)	Doctorate	273,077	270,360	264,995
Sciences (excluding health fields)	Master's	42,279	41,780	41,641
Engineering	Total of All U.S. Institutions	107,199	103,223	101,008
Engineering	Doctorate	99,917	96,445	94,481
Engineering	Master's	7,282	6,778	6,527
Mathematical and Computer Sciences	Total of All U.S. Institutions	51,941	52,607	52,769

Table 8.4. Total Master's and Doctorate Degrees Granted in
Selected Engineering and Science Disciplines 1995–1997* *(cont'd)*

Graduate Student Survey —Academic Discipline	Highest Degree Granted	1995	1996	1997
Mathematical and Computer Sciences	Doctorate	44,678	45,106	44,969
Mathematical and Computer Sciences	Master's	7,263	7,501	7,800
Computer Science	Total of All U.S. Institutions	33,432	34,592	36,010
Computer Science	Doctorate	28,270	29,111	30,106
Computer Science	Master's	5,162	5,481	5,904
Mathematics and Applied Mathematics	Total of All U.S. Institutions	15,400	14,970	14,078
Mathematics and Applied Mathematics	Doctorate	13,380	13,003	12,232
Mathematics and Applied Mathematics	Master's	2,020	1,967	1,846
Statistics	Total of All U.S. Institutions	3,109	3,045	2,681
Statistics	Doctorate	3,028	2,992	2,631
Statistics	Master's	81	53	50

*Another display of data collected by the National Science Foundation, showing the number of masters and doctorate degrees granted in selected engineering and science disciplines throughout the United States.

Source: NSF WebCASPAR Database System. http://caspar.nsf.gov/.

Table 8.5. Selected Science and Engineering Postdoctoral Appointees
and Other Nonfaculty Doctoral Research Staff in Doctorate-Granting
Institutions by Gender, 1997*

Field	Postdoctoral Appointees			Other Nonfaculty Doctoral Research Staff		
	Total	Men	Women	Total	Men	Women
Total, All Surveyed Fields	37,928	26,120	11,808	6,674	4,709	1,965
Total, Science and Engineering Fields	26,806	19,027	7,779	5,212	3,862	1,350
Sciences, Total	23,868	16,433	7,435	4,362	3,127	1,235
Mathematical Sciences	303	263	40	92	81	11
Computer Sciences	315	269	46	87	71	16
Engineering, Total	2,938	2,594	344	850	735	115
Electrical Engineering	505	465	40	168	150	18
Engineering Science	115	94	21	50	42	8

Table 8.5. Selected Science and Engineering Postdoctoral Appointees and Other Nonfaculty Doctoral Research Staff in Doctorate-Granting Institutions by Gender, 1997* *(cont'd)*

Field	Postdoctoral Appointees			Other Nonfaculty Doctoral Research Staff		
	Total	Men	Women	Total	Men	Women
Industrial Engineering/ Manufacturing Engineering	27	27	0	7	3	4
Mechanical Engineering	436	409	27	109	96	13
Metallurgical/Materials Engineering	464	402	62	84	68	16

*These data demonstrate the disparity between numbers of males versus females appointed to postdoctoral science and engineering positions in U.S. university programs for the academic year 1997.

Source: National Science Foundation (NSF)/SRS Survey of Graduate Students and Postdoctorates in Science and Engineering. http://caspar.nsf.gov/.

Table 8.6. Graduate Students in Science, Engineering, and Health Fields in All Institutions, by Field, 1990-1997*

Sciences

Year	Total, Science and Engineering	Sciences, Total	Physical Sciences	Earth, Atmospheric, & Ocean Sciences	Mathe-matical Sciences	Computer Sciences	Agricul-tural Sciences	Biological Sciences	Psychology	Social Sciences
1990	397,135	289,510	34,075	13,984	19,774	34,257	11,316	49,989	48,167	77,948
1991	412,697	299,121	34,710	14,480	19,952	34,610	11,506	51,778	51,343	80,742
1992	430,644	312,609	35,348	15,347	20,355	36,293	11,827	54,177	53,484	85,778
1993	435,886	319,028	35,318	15,805	20,000	36,189	11,914	56,452	54,557	88,793
1994	431,251	318,228	34,449	16,042	19,579	34,128	12,199	58,143	54,554	89,134
1995	422,555	315,356	33,417	15,805	18,509	33,432	12,367	58,736	53,641	89,449
1996	415,363	312,140	32,355	15,280	18,015	34,592	11,914	58,128	53,209	88,647
1997	407,644	306,636	31,108	14,644	16,759	36,010	11,810	57,135	53,142	86,028

Engineering

Year	Engineering, Total	Aerospace Engineering	Chemical Engineering	Civil Engi-neering	Electrical Engineering	Industrial Engineering	Mechanical Engineering	Metallurgical & Materials Engineering	Other Engineering	Health Fields
1990	107,625	3,934	6,735	15,542	33,722	11,248	16,879	4,941	14,624	55,043
1991	113,576	4,120	7,127	17,398	35,182	12,676	17,730	5,160	14,183	58,565
1992	118,035	4,036	7,397	19,572	36,460	13,525	18,637	5,512	12,896	62,988
1993	116,858	3,940	7,516	19,583	35,314	13,596	18,477	5,363	13,069	68,563
1994	113,023	3,715	7,608	19,925	33,050	13,661	17,761	5,191	12,112	73,291
1995	107,199	3,343	7,424	19,218	30,747	13,143	16,363	4,920	12,041	77,177
1996	103,223	3,208	7,373	18,528	29,736	12,399	15,509	4,713	11,757	78,856
1997	101,008	3,083	7,247	17,033	30,617	11,725	15,044	4,649	11,610	79,460

*These NSF data show the trends in numbers of graduate students in all science and engineering programs in U.S. university programs for the academic years 1990-1997.

Source: National Science Foundation (NSF)/SRS Survey of Graduate Students and Postdoctorates in Science and Engineering. http://caspar.nsf.gov/.

PROJECTED AREAS OF JOB GROWTH

Table 8.7. Fastest Growing Occupations, 1996–2006*

Occupation	Year 1996	Year 2006	Number Change	% Change	Education and Training
Database Administrators, Computer Support Specialists, and All Other Computer Scientists	212	461	249	118	Bachelor's Degree
Computer Engineers	216	451	235	109	Bachelor's Degree
Systems Analysts	506	1,025	520	103	Bachelor's Degree
Personal and Home Care Aides	202	374	171	85	Short-Term on-the-Job Training
Physical and Corrective Therapy Assistants and Aides	84	151	66	79	Moderate-Term on-the-Job Training
Home Health Aides	495	873	378	76	Short-Term on-the-Job Training
Medical Assistants	225	391	166	74	Moderate-Term on-the-Job Training
Desktop Publishing Specialists	30	53	22	74	Long-Term on-the-Job Training
Physical Therapists	115	196	81	71	Bachelor's Degree
Occupational Therapy Assistants and Aides	16	26	11	69	Moderate-Term on-the-Job Training
Paralegals	113	189	76	68	Associate's Degree
Occupational Therapists	57	95	38	66	Bachelor's Degree
Teachers, Special Education	407	648	241	59	Bachelor's Degree
Human Services Workers	178	276	98	55	Moderate on-the-Job Training

Table 8.7. Fastest Growing Occupations, 1996–2006* *(cont'd)*

Data Processing Equipment Repairers	80	121	42	52	Post-secondary Vocational Training
Medical Records Technicians	87	132	44	51	Associate's Degree
Speech-Language Pathologists and Audiologists	87	131	44	51	Master's Degree
Dental Hygienists	133	197	64	48	Associate's Degree
Recreation Attendants	288	426	138	48	Short on-the-Job Training
Physician Assistants	64	93	30	47	Bachelor's Degree
Respiratory Therapists	82	119	37	46	Associate's Degree
Adjustment Clerks	401	584	183	46	Short on-the-Job Training
Engineering, Science, and Computer Systems Managers	343	498	155	45	Work Experience Plus Bachelor's or Higher Degree

*Occupations are ranked based on projected percent changes in employment from year 1996 to 2006; numerical employment changes are in thousands of jobs.

Source: Occupational employment projections to 2006,"Monthly Labor Review." U.S. Bureau of Labor Statistics. Also available via the "Occupational Outlook Quarterly" at www.bls.gov/ocohome.htm.

LIST OF THE WORLD'S MOST POWERFUL COMPUTING SITES

The list (the total of which is substantially longer than what is excerpted here) is included with the permission of its author, Gunter Ahrendt (gunter@gapcon.com). Ahrendt maintains an updated copy of the listing of the most powerful computers at www.gapcon.com/list.html. The 20 sites that follow represent about 10% of the list available on January 7, 2000. The criteria for inclusion is peak performance of 95.6 GFLOPS. (A GFLOPS is a unit abbreviation equal to one thousand million floating point operations per second). Judging computing power via the GFLOPS standard is the most common method for comparing supercomputer performance. (See GFLOPS in Glossary for more information.)

The information contained here includes ranking, aggregate GFLOPS computing capacity, the date of testing/reporting, the name of the site where the computer(s) are located, the acronym for that site, its uniform resource locator (URL), the names of the individual machines, and the peak GFLOPS performance achieved with each. The asterisks (*) indicate multiple machines; for example, 3* Cray means 3 Cray machines.

Table 8.8. World's Most Powerful Supercomputing Sites

1) 6427.12 - (03-DEC-1999) [**LLNL**]
   Lawrence Livermore National Lab, Livermore, California, US

1)	IBM RS/6000 SP-604e/5856 [-3Q00]	3905.95
2)	Compaq AlphaServer SC1024/667	1366.02
3)	IBM RS/6000 SP-604e/1408 [-3Q00]	939.14
4)	IBM RS/6000 SP-POWER3/222-128 [-2Q00]	113.67
5)	24 * DEC AlphaServer 4100/533-4	102.34
6)	IBM RS/6000 SP-POWER3/300-8192 [+2Q00,-2Q02]	9830.4
7)	IBM T30/10240 [+2Q01,-3Q03]	30000

2) 5965.13 - (04-NOV-1999) [**S390**]
   IBM, Poughkeepsie, New York, US

1)	IBM RS/6000 SP-POWER3/300-4096 [-1Q00]	4915.2
2)	IBM RS/6000 SP-POWER3/200-512	409.6
3)	IBM RS/6000 SP-604e/488	323.91
4)	IBM RS/6000 SP-P2SC/309-256	316.42
5)	IBM RS/6000 SP-POWER3/300-8192 [+1Q00,-2Q00]	9830.4

3) 5253.48 - (24-NOV-1999) [**NSA**]
   National Security Agency, Fort Meade, Maryland, US

1)	Cray T3E-1200 LC1084	1300.8
2)	Cray T3E-900 LC1328	1195.2
3)	Cray SV1-18/576 [-3Q02]	691.2
4)	SGI 2800/250-1152	576

Table 8.8. World's Most Powerful Supercomputing Sites *(cont'd)*

5)		Cray T3E-1200 LC284	340.8
6)	3 *	Cray T932/321024	174
7)	4 *	Sun Ultra HPC 10000-64336	170.48
8)		Cray T3E-750 LC220	165
9)		Paracel FDF3-8T/2240	160
10)		Paracel FDF3-8T/2240	160
11)		Paracel FDF3-8T/2240	160
12)		Paracel FDF3-8T/2240	160
13)		Cray SV2-18/576 [+3Q02]	1382.4

4) 4992 - (10-JUN-1999) [**RSC**]
Raytheon Systems, Garland, Texas, US

1)	SGI 2800/300-8320 [-4Q02]	4992
2)	SGI 2800/300-11776 [+4Q02]	7065.6

5) 4833.04 - (11-NOV-1999) [**LANL**]
Los Alamos National Lab, Los Alamos, New Mexico, US

1)		SGI 2800/250-6144 [-3Q00]	3072
2)		SGI 2800/250-2200 [-3Q00]	1100
3)	256 *	Atipa ATserver 6000/500-4	512
4)	140 *	DEC AlphaPC/533	149.24
5)		Compaq/SGI/Sun T30/10240 [+2Q01,-3Q03]	30000
6)		Compaq/IBM/SGI/Sun T100 [+1Q04]	100000

6) 4659.4 - (11-NOV-1999) [**SANDIA**]
Sandia National Labs, Albuquerque, New Mexico, US

1)		Intel ASCI Red/P2-333/9632	3207
2)	800 *	Compaq XP1000/500-450*PC/500-16*DS20/500-6*	
		1200/533	1272.4
3)	12 *	DEC AlphaServer 8400/625-12	180

7) 2265.6 - (11-NOV-1999) [**CRAY-MN**]
SGI, Eagan, Minnesota, US

1)	Cray T3E-1200E LC1536	1843.2
2)	SGI 2800/300-384	230.4
3)	SGI 2800/250-384	192

8) 1560 - (24-NOV-1999) [**METO**]
Meteorological Office, Bracknell, England

1)	Cray T3E-900 LC880 [-4Q03]	792
2)	Cray T3E-1200E LC640 [-4Q03]	768

9) 1510.4 - (15-JUN-1999) [**U-TOKYO**]
Uni of Tokyo, Minato-ku, Tokyo, Japan

1)	Hitachi SR8000/128	1024
2)	Hitachi SR2201/1280	384
3)	SGI 2800/200-256	102.4

Table 8.8. World's Most Powerful Supercomputing Sites *(cont'd)*

10) 1328 - (11-NOV-1999) **[SCHWAB]**
   Charles Schwab, New York, New York, US
      1)   IBM RS/6000 SP-604e/2000   1328

11) 1326.88 - (17-DEC-1999) **[ERDC]**
   US Army Engineering & Research Development
     Center, Vicksburg, Mississippi, US
      1)   Cray T3E-1200E LC544   652.8
      2)   IBM RS/6000 SP-POWER3/222-512   454.66
      3)   IBM RS/6000 SP-P2SC/135-256/160-126 [-4Q01]   219.42

12) 1234 - (29-DEC-1999) **[C]**
   Uni of Tokyo, Meguro-ku, Tokyo, Japan
      1)   U-Tokyo GRAPE-4/1792   1080
      2)   U-Tokyo GRAPE-5/32 [-2Q00]   154
      3)   U-Tokyo GRAPE-6/16 [+1Q00,-4Q00]   576
      4)   U-Tokyo GRAPE-5/128 [+2Q00]   616
      5)   U-Tokyo GRAPE-6/2048 [+4Q00,-4Q01]   73728
      6)   U-Tokyo GRAPE-6/4096 [+4Q01]   147456

13) 1200 - (03-DEC-1999) **[CELERA]**
   Celera, Rockville, Maryland, US
      1)   Compaq AlphaServer SC1200/500   1200

14) 1186.17 - (07-JAN-2000) **[SDSC]**
   San Diego Supercomputer Center, San Diego, California, US
      1)   IBM RS/6000 SP-POWER3/222-1152   1022.97
      2)   Cray T3E-600 LC272   163.2

15) 1112.8 - (09-DEC-1999) **[LBL]**
   Lawrence Berkeley National Lab, Berkeley Hills, California, US
      1)   Cray T3E-900 LC696   626.4
      2)   IBM RS/6000 SP-POWER3/200-608 [-4Q00]   486.4
      3)   IBM RS/6000 SP-POWER3/300-2731 [+4Q00]   3277.2

16) 1098.4 - (24-NOV-1999) **[NAVO]**
   Naval Oceanographic Office, NASA Stennis Space Center, Bay Saint Louis,
     Mississippi, US
      1)   Cray T3E-900 LC1088   979.2
      2)   SGI 2800/250-136/200-128   119.2

17) 1008 - (10-AUG-1999) **[PIXAR]**
   Pixar, Point Richmond, California, US
      1) 120 * Sun Ultra 4500-14300   1008

Table 8.8. World's Most Powerful Supercomputing Sites *(cont'd)*

18) 979.2 - (24-NOV-1999) **[DWD]**
German Weather Service, Offenbach, Germany
   1)   Cray T3E-1200E LC816   979.2

19) 974.4 - (11-NOV-1999) **[FZ-JUELICH]**
Research Center Juelich, Juelich, Germany
   1)   Cray T3E-1200 LC540   648
   2)   Cray T3E-600 LC544   326.4

20) 972.8 - (03-DEC-1999) **[TACC]**
Tsukuba Advanced Computing Center, Tsukuba, Ibaraki, Japan
   1)   Hitachi SR8000/64   512
   2) 256 * Compaq AlphaPC/500   256
   3)   IBM RS/6000 SP-POWER3/200-256 204.8

Source: Ahrendt, Gunter. 2000. "List of the World's Most Powerful Computing Sites." www.gapcon.com/list.html. (3 February 2000).

CHAPTER NINE
Organizations and Associations

The organizations included in the following pages are a representative sample of some of the most respected international and national groups dedicated to the promulgation of computer science, computational science, and computer engineering information. Many are societies that have been founded to help professionals expand and enhance knowledge that will further careers and improve the field to which they are dedicated. After listing the name, address, phone number, and URL for each organization, a brief description of the goal or particular objectives is included.

Alliance for Telecommunications Industry Solutions (ATIS)
1200 G Street, NW, Suite 500
Washington, DC 20005
(202) 628-6380
(202) 393-5453, fax
www.atis.org

This industry group of over 2,000 telecommunications experts promotes "the timely resolution of national and international issues involving telecommunications standards and the development of operational guidelines. ATIS will initiate and maintain flexible, open industry forums to address technical and operational issues affecting the nation's telecommunications facilities and services and the development of innovative technologies."

American Computer Scientists Association (ACSA)
11 Commerce Drive, 3rd Floor
Cranford, NJ 07016-3531
(908) 272-0016
www.acsa2000.net

ACSA is a nonprofit, charitable research and education organization formed from the American Computing and the American Certified Computer Consultants Associations, and the American Certified Computer Scientists Association. Anyone is eligible to join for free.

American Electronics Association (AEA)
601 Pennsylvania Ave, NW
North Building, Suite 600
Washington, DC 20004
(202) 682-9110
(202) 682-9111, fax
or
5201 Great America Parkway, Suite 520
Santa Clara, CA 95054
(408) 987-4280
(408) 986-1247, fax
www.aeanet.org

AEA claims to be "the largest and most effective trade association of the technology industry." Founded in 1943, the group now maintains a membership of 3,000 electronics associated companies. The following industry segments are represented among this advocacy organization: software, semiconductors/components, computers, telecommunications, business-related services, instrumentation, manufacturing, medical equipment, and aerospace/defense. The organization makes its position known on the important issues involved in crucial aspects of software, electronics, and information technology.

American National Standards Institute (ANSI)
11 West 42nd Street
New York, NY 10036
(212) 642-4900
(212) 398-0023, fax
web.ansi.org

Founded in 1918 by five engineering societies and three government agencies, ANSI is a voluntary private sector agency that administrates and coordinates technical standards. Its stated primary goal is "the enhancement of global competitiveness of U.S. business and the American quality of life by promoting and facilitating voluntary consensus standards and conformity assessment systems and promoting their integrity." The institute represents the wishes of almost 1,400 members. The organization maintains an extensive reference library at www.ansi.org/public/ref_lib.html.

American Society for Information Science (ASIS)
8720 Georgia Avenue, Suite 501
Silver Spring, MD 20910
(301) 495-0900

(301) 495-0810, fax

www.asis.org

ASIS is an organization founded in 1937 that works in an interdisciplinary manner to sustain new developments in information technology. ASIS members number some 4,000 information specialists from such fields as computer science, linguistics, management, library science, engineering, law, medicine, chemistry, and education; it is basically a group of individuals who share a common interest in improving the ways society stores, retrieves, analyzes, manages, archives, and disseminates information coming together for mutual benefit.

ASIS publications include the *Journal of the American Society for Information Science* (JASIS), *Bulletin of the American Society for Information Science*, and *Proceedings of ASIS Meetings*.

Association for Computing Machinery (ACM)
One Astor Plaza
1515 Broadway
New York, NY 10036
(212) 869-7440
(212) 944-1318, fax
www.acm.org

Boasting a membership of over 80,000, ACM is the premier computer association in the world. The group was founded in 1947 by John Mauchly, coinventor of the ENIAC, the first general purpose computer. The association is dedicated to advancing the art, science, engineering, and application of information technology. Publications of the ACM include *Communications of the ACM*, *Crossroads: The International ACM Student Magazine*, and *Journal of the ACM*.

Association for Information Systems (AIS)
P.O. Box 2712
Atlanta, GA 30301-2712
(404) 651-0258
(404) 651-4938, fax
www.aisnet.org

This information professionals organization's purpose is to serve as "the premier global organization for academicians specializing in Information Systems." AIS supports several international conferences including the AIS-PERT series, the Annual Conference of the Southern Chapter of the Association for Information Systems (SAIS), ECIS '99, the 7th European Conference on Information Systems in Copenhagen, Denmark, and BITWorld.

Publications of the AIS include *Information Systems*, *Journal of Organizational Computing*, and *MIS Quarterly*.

Association for Women in Computing (AWC)
41 Sutter Street, Suite 1006
San Francisco, CA 94104
(415) 905-4663
www.awc-hq.org

This group was started in 1978 to promote communication among women in computing; further the professional development and advancement of women in computing; and promote the education of women of all ages in computing. The AWC sponsors meetings, publishes both print and electronic newsletters, and encourages networking. AWC publishes two informational products: the *AWC Source* and the *Association for Women in Computing NewsBytes*.

Association for Women in Science
1200 New York Avenue, Suite 650
Washington, DC 20005
(202) 326-8940
(202) 326-8960, fax
www.awis.org

AWIS is a nonprofit organization. It was established in 1971 to achieve "equity and full participation for women in science, mathematics, engineering, and technology." Boasting a membership of over 5,000 members from all disciplines of science and engineering, one of AWIS's key goals, which it undertakes through a network of 76 local chapters, is the fostering of careers of women science professionals. An important element of this work is accomplished by encouraging young women to take up science careers by sponsoring educational activities in schools and within communities. Publications of the Association for Women in Science include the bimonthly *AWIS Magazine*.

Austrian Computer Society
Wollzeile 1-3, 1. Stock
A-1010 Wien
Vienna, Austria
(43) 1-5120235
(43) 1-5120235 9, fax
www.ocg.or.at/ENGL/general.html (English version)

The objective of the Austrian Computer Society "is the comprehensive and interdisciplinary promotion of information processing, with due regard to its effects on man and society." To accomplish this goal, the European group serves as an umbrella organization in Austria for institutions involved in information processing. The organization is an affiliate of both the Association for Computing Machinery (ACM) and of the Institute of Electrical and Electronics Engineers (IEEE) Computer Society, and it has established 22 working groups to deal with aspects of information technology.

Bell Laboratories Information Sciences Research Center
Bell Laboratories
600 Mountain Avenue
Murray Hill, NJ 07974
www.bell-labs.com/org/1123/

Part of the Computing and Mathematical Sciences Research Division of Bell Laboratories, the Bell Laboratories Information Sciences Research Center is under the direction of Avi Silberschatz. His leadership provides direction on research projects that affect issues in computer science related to information management systems and the structure of telecommunication software and systems. Research topics include the design and analysis of algorithms, data models, performance models, database systems, operating systems, networking, and distributed systems; and the security of information and communication at all levels, including data transmission security, operating system security, and database system security.

Benton Foundation
1634 Eye Street, NW, 12th Floor
Washington, DC 20006
(202) 638-5770
(202) 638-5771, fax
www.benton.org

According to their mission statement, the Benton Foundation "works to realize the social benefits made possible by the public interest use of communications. Bridging the worlds of philanthropy, public policy, and community action, Benton seeks to shape the emerging communications environment and to demonstrate the value of communications for solving social problems." It provides updated alerts and informational briefings on salient areas of communications policy directed to those who want to keep abreast of government and corporate policies that could impact the egalitarian nature of the developing infrastructure.

Black Data Processing Associates (BDPA)
8401 Corporate Drive, Suite 405
Landover, MD 20785
(301) 429-2702
(301) 429-2710, fax
www.bdpa.org

Since 1975, BDPA has served as a valuable liaison between the information technology industry and African American communities. The organization supports the efforts of more than 40 local chapters that offer services including career counseling, technological assistance, networking opportunities, workshops, and computer competitions.

British Computer Society
1 Sanford Street
Swindon, Wiltshire SN1 1HJ
England
(44) 01793 417417
(44) 01793 480270, fax
www.bcs.org.uk

The British Computer Society is the only chartered professional body representing computer professionals in England. It was established to monitor individual professional competence and integrity; define standards for professional conduct; advise the UK Government and its agencies on IS-related matters; examine topical information systems issues such as software certification, intellectual property rights, pornography, and the year 2000 (Y2K) problem; and to set standards for education and training through the inspection of school coursework. It publishes *The Computer Journal*.

Canadian Information Processing Society (CIPS)
One Yonge Street, Suite 2401
Toronto, Ontario M5E 1E5
(416) 861-2477
(416) 368-9972, fax
www.cips.ca

According to home page of this national Canadian information technology (IT) support organization, its mission is "to define and foster the IT profession, to encourage and support the IT practitioner, and to advance the theory and practice of IT, while safeguarding the public interest." It strives to do this through the advancement of theory and practice of information technology in Canada, via the promotion and exchange of information about IT, to monitor the competence of Canadian IT professionals, and to inform the public regarding IT issues. It produces the *CIPS National Newsletter.*

Center on Information Technology Accommodation (CITA)
General Services Administration
Center for Information Technology Accommodation
KBA Room 1234
18th & F Street, NW, Room 1234
Washington, DC 20405
(202) 501-4906
(202) 501-2010, tdd
(202) 501-6269, fax
www.itpolicy.gsa.gov/cita/

CITA is a clearinghouse for information and recommendations regarding ways to make information systems accessible to all users. It is a federal entity and works under the auspices of the General Services Administration. CITA publishes *Managing Information Resources for Accessibility* which is available online at www.itpolicy.gsa.gov/cita/front.htm.

CERT Coordination Center (CERT/CC)
Software Engineering Institute
Carnegie Mellon University
Pittsburgh, PA 15213-3890
(412) 268-7090
(412) 268-6989, fax
www.cert.org

As part of the Survivable Systems Initiative at the Software Engineering Institute at Carnegie Mellon University, this center was originally created by the Defense Applied Research Projects Agency (DARPA) in December 1988. The center was established in response to what has been named "the Morris Worm incident," a virus responsible for taking down approximately 10% of all computers connected to the Internet.

Today the center is the chief repository and alert facility for issues regarding Internet security vulnerabilities, prevention of vulnerabilities, system security improvement, and survivability of large-scale networks. Besides widely distributing CERT alerts when incidents of malicious hacker activity (mainly viruses) are detected, the group also publishes some important papers that are available on their Web site.

These include *Handbook for Computer Security Incident Response Teams, rlogin(1): The Untold Story, Security of the Internet*, and *Report to the President's Commission on Critical Infrastructure Protection.*

Computer Professionals for Social Responsibility (CPSR)
P.O. Box 717
Palo Alto, CA 94302
(650) 322-3778
(650) 322-4748, fax
www.cpsr.org

CPSR was founded in 1981 by computer scientists who were worried about the use of computers in nuclear weapons systems. Today, "CPSR members provide the public and policymakers with realistic assessments of the power, promise, and limitations of computer technology. As concerned citizens, we direct public attention to critical choices concerning the applications of computing and how those choices affect society." General areas of concern include the national information infrastructure, civil liberties and privacy, computers in the workplace, technology policy and human needs, and reliability and risk of computer-based systems. CPSR regularly publishes the *CPSR Newsletter.*

Computing Research Association (CRA)
1100 Seventeenth Street, NW, Suite 507
Washington, DC 20036-4632
(202) 234-2111
(202) 667-1066, fax
www.cra.org

Representing more than 180 North American academic departments of computer science and engineering, and laboratories and centers in industry and government, the group's mission is to strengthen research and education in the computing fields. CRA also strives to increase opportunities for women and minorities in the computer science field while improving public understanding of the importance of computing and computing research. The organization publishes the newsletters *CRA* and *CRA Online* (www.cra.org/CRN).

Computing Research Association (CRA) Committee on the Status of Women in Computer Science and Engineering
100 Seventeenth Street, NW, Suite 507
Washington, DC 20036-4632
(202) 234-2111
(202) 667-1066, fax
www.cra.org/Activities/craw/

According to their home page, "the goal of the CRA Committee on the Status of Women in Computer Science and Engineering (CRA-W) is to take positive action to increase the number of women participating in Computer Science and Engineering (CSE) research and education at all levels. The organization makes the careers booklet "Women in Computer Science" available online for free at www.sdsc.edu/CRAW/careers/.

Conseil European pour la Recherche Nucleaire (CERN)
European Laboratory for Particle Physics
CH - 1211
Geneva 23 Switzerland
(41) 22 767 6111
(41) 22 767 6555, fax
www.cern.ch/WebOffice

The European Laboratory for Particle Physics is the division of CERN that was responsible for the first protocols and first deployment of the Web software, which were directed by Tim Berners-Lee. The lab continues to be the center of much of the activity that impacts technical specifications and international standards even though the WWW Consortium has moved to MIT.

Debian Project
www.debian.org/

The project was started to develop and distribute the free (Open Source) operating system based on the Linux kernel. "The kernel is the most fundamental program on the computer, does all the basic housekeeping and lets you start other programs. . . . A large part of the basic tools that fill out the operating system come from GNU, which are also free." Debian also has 1,500 other packages available for free.

Defense Advanced Research Projects Agency (DARPA)
3701 North Fairfax Drive
Arlington, VA 22203-1714
(703) 696-0104
(703) 528-3655, fax
www.arpa.mil

DARPA is the central research and development organization for the U.S. Department of Defense. It was the agency's goal in the early 1960s to build a nonhierarchical communications network that could withstand a nuclear attack. The result of that effort was the ARPANet, the precursor of today's Internet. Currently, DARPA still acts as the locus of military "research and technology where risk and payoff are both very high and where success may provide dramatic advances for traditional military roles and missions and dual-use applications."

Electronic Frontier Foundation (EFF)
1550 Bryant Street, Suite 725
San Francisco, CA 94103-4832
(415) 436-9333
(415) 436-9993, fax
www.eff.org

Founded in 1990 as a nonprofit, public interest organization, the Electronic Frontier Foundation is one of the leading civil liberties groups devoted to ensuring that the Internet remains a medium for the free exchange of ideas. As EFF literature says, "new digital networks offer a tremendous potential to empower individuals in an ever-overpowering world. However, these communications networks are also the subject of significant debate concerning governance and jurisdiction." The EFF publishes the newsletter *EFFector* which can be found at www.eff.org/effector/.

Electronic Privacy Information Center (EPIC)
666 Pennsylvania Avenue, SE, Suite 301
Washington, DC 20003
(202) 544-9240
(202) 547-5482, fax
www.epic.org

EPIC is a public interest research center in Washington, DC. It was established in
1994 to focus public attention on emerging civil liberties issues and to protect
privacy, First Amendment rights, and other constitutional values. EPIC is a project
of the Fund for Constitutional Government. EPIC works in association with
Privacy International, an international human rights group based in London,
England, and is also a member of the Global Internet Liberty Campaign, the
Internet Free Expression Alliance, and the Internet Privacy Coalition. The organization publishes an award-winning newsletter, EPIC Alert, which is available online at
www.epic.org/alert/.

Fibre Channel Industry Association
2570 West El Camino Real, Suite 304
Mountain View, CA 94040-1313
(650) 949-6730
(650) 949-6735, fax
www.fibrechannel.com

Fibre Channel Industry Association (FCIA) is dedicated to providing information and
a showcase for products that adhere to the new fibre channel standard. This high
speed "gigabit interconnect technology allows concurrent communications among
workstations, mainframes, servers, data storage systems, and other peripherals using
SCSI and IP protocols. It provides interconnect systems for multiple topologies that
can scale to a total system bandwidth on the order of a terabit per second. Fibre
Channel delivers a new level of reliability and throughput. Switches, hubs, storage
systems, storage devices, and adapters are among the products that are on the market
today, providing the ability to implement a total system solution."

Free Software Foundation (FSF)
59 Temple Place, Suite 330
Boston, MA 02111
www.gnu.org

The Free Software Foundation (FSF) is the catalyst behind the development of the
GNU (Gnu's not Unix) operating system, a free replacement for UNIX. GNU was
developed in 1983 by FSF in an effort to eliminate restrictions on copying,
redistributing, understanding, and modifying computer programs. The foundation
distributes free copies of GNU and new software that does not have to use commercial "proprietary" code. The goal of the FSF is to bring back the camaraderie
that existed among programmers in the earliest days of computer code development.

Institute for Certification of Computing Professionals (ICCP)
2200 East Devon Avenue, Suite 247
Des Plaines, IL 60018
(847) 299-4227
(847) 299-4280, fax
www.iccp.org

The ICCP began in 1973 for the express purpose of creating a certification/testing regime for individuals wanting to bypass or enhance college or university computer science programs. It has the support of nine constituent societies and 18 affiliate societies that help to define and update information technology standards. The institute offers two professional designations—Certified Computing Professional (CCP) and Associate Computing Professional (ACP). Recently the organization has established a nonprofit educational foundation to further the cause of professional certification.

Institute of Electrical and Electronics Engineers (IEEE)
445 Hoes Lane
Piscataway, NJ 08855-1331
(908) 981-0060
(908) 981-0345, fax
www.ieee.org

Founded in 1884, with a membership today of over 320,000 in nearly 150 countries, the IEEE is the world's largest technical professional society. The society's mission is to support the application of "electrotechnology and allied sciences for the benefit of humanity and the advancement of the profession."

The IEEE publishes almost one-quarter of the world's technical papers in electronics and computer engineering. For more information on these publications, see www.ieee.org/organizations/pab.html.

Institute of Electrical and Electronics Engineers (IEEE) Communications Society
305 E. 47th Street
New York, NY 10017
(202) 371-0101
(202) 728-9614, fax
www.comsoc.org

This subgroup of the IEEE was founded in 1952 to provide a venue that would "foster original work in all aspects of communications science, engineering, and technology" and to "encourage the development of applications that use signals to transfer voice, data, image, and/or video information between locations."

Institute of Electrical and Electronics Engineers (IEEE) Computer Society
1730 Massachusetts Avenue, NW
Washington, DC 20036-1992
(202) 371-0101
(202) 728-9614, fax
computer.org

Founded in 1946 under the auspices of the Institute of Electrical and Electronics Engineers, the Computer Society is one of the main organizations for professionals and students who seek to keep abreast of current discoveries in computer science.

"The oldest and largest association of computing professionals in the world, the Computer Society serves more than 90,000 members from its headquarters in Washington, DC, as well as offices in California, Tokyo, and Brussels." The group publishes several influential journals including *Computer*, *IT Professional*, and *IEEE Internet Computing*.

International Society for Computers and Their Applications (ISCA)
975 Walnut Street, Suite 132
Cary, NC 27511-4216
(919) 467-5559
(919) 467-3430, fax
www.isca-hq.org

The International Society for Computers and Their Applications, Inc., promotes the advancement of science and engineering in the area of computers and their applications, and disseminates this technology throughout the world.

ISCA is a not-for-profit society which publishes the *International Journal of Computers and Their Applications*.

Internet Assigned Numbers Authority (IANA)
P.O. Box 12607
Marina del Rey, CA 90292-3607
(310) 822-1511
(310) 823-6714, fax
www.iana.org

IANA is the brainchild of Internet legend Jon Postel who managed its operations from the onset of the Internet. IANA oversees the IP addresses and naming registry of the Internet, performing a critical and necessary central coordinating function. It will soon be replaced by a nongovernmentally sponsored international group called the Internet Corporation for Assigned Names and Numbers (ICANN). (See next entry for details.)

Internet Corporation for Assigned Names and Numbers (ICANN)
339 La Cuesta Drive
Portola Valley, CA 94028
(650) 854-2108
(650) 854-8134, fax
www.icann.org

ICANN is the new nonprofit corporation that was formed to take over responsibility for the IP address space allocation, protocol parameter assignment, domain name system management, and root server system management functions now performed under U.S. government contract by IANA and other entities. The intent, for which the federal government has given tentative approval, is for this organization to take on the job of assigning unique IP addresses (domain names) for Internet sites.

Internet Engineering Task Force (IETF)
c/o Corporation for National Research Initiatives
1895 Preston White Drive, Suite 100
Reston, VA 22091
(703) 620-8990

(703) 758-5913, fax
www.ietf.org

The IETF is an international group of network designers, operators, vendors, and researchers concerned with the evolution of the Internet architecture and its smooth operation. This task force does its work on the technical considerations of maintaining the network of networks via its working groups, which are organized into several areas, such as routing, transport, security, etc.

Internet Society
12020 Sunrise Valley Drive, Suite 210
Reston, VA 20191-3429
(703) 326-9880
(703) 326-9881, fax
info.isoc.org

Representing more than 100 organizations and 6,000 individual members worldwide, the Internet Society is an international educational organization for global coordination for the Internet and its networking technologies and protocols. Since 1992 the society has acted "not only as a global clearinghouse for Internet information and education but also as a facilitator and coordinator of Internet-related initiatives around the world. Through its annual International Networking (INET) conference and other sponsored events, developing-country training workshops, tutorials, statistical and market research, publications, public policy and trade activities, regional and local chapters, standardization activities, committees and an international secretariat, the Internet Society serves the needs of the growing global Internet community."

John von Neumann Computer Society
H-1054 Budapest, Báthori u. 16.
(36-1) 332-9349
(36-1) 332-9390
(36-1) 331-8140, fax
www.njszt.iif.hu/1_main.htm (English version)

Founded in 1968, this 3,500+ member organization represents some 200 institutions in and around Hungary. Goals of the group are to promote the study of computer technology by informing professionals on the latest practices and developments of their peers and by promulgating computer information among the user community. The society is aptly named for von Neuman (born Neumann János in Budapest in 1903) one of the world's most famous computer scientists for his work on ENIAC, the first general purpose computer.

Linux International
80 Amherst Street
Amherst, NH 03031-3032
www.li.org/index.shtml

Linux International is an international nonprofit organization located in the United States. It has been established to promulgate information about Linux regarding how it might benefit business and personal users.

Linux Online
www.linux.org/

This group is the main body responsible for propagating the Linux operating system. At their online site, anything that has to do with Linux is either directly available or connected via hyperlink.

National Computational Science Alliance
152 Computing Applications Building
605 East Springfield Avenue
Champaign, IL 61820-5518
(217) 244-0072
(217) 265-0460, fax
www.ncsa.uiuc.edu/alliance

This is the division of the University of Illinois' National Center for Supercomputing Applications that has lead responsibility for creating "the Grid." This experimental, interconnected, network of computers, applications, and new protocols may prove to be the computing environment of the future.

National Coordination Office for High Performance Computing and Communications
4201 Wilson Boulevard, Suite 665
Arlington, VA 22230
(703) 306-HPCC
(703) 306-4727, fax
www.hpcc.gov

The mission of the National Coordination Office (NCO) for Computing, Information, and Communications (CIC) is to support the Committee on Computing, Information, and Communications (CCIC), which reports to the National Science and Technology Council. The NCO provides technical and administrative support to the CCIC's Computing, Information, and Communications' (CIC) R & D Subcommittee, and provides administrative support to the Federal Networking Council (FNC), the Applications Council, and the Technology Policy Subcommittee (TPS), all of which report to the CCIC.

The NCO coordinates multiagency CIC R&D activities in computing, information, and communications. These activities include the preparation of planning, budget, and assessment documents, including the CIC R&D Annual Report, which is a supplement to the president's budget and is required by law, and the CIC R&D Implementation Plan; development of multiagency activities; and information exchanges. The subcommittee has five working groups that address the following specific programmatic objectives: High End Computing and Computation; Large Scale Networking; High Confidence Systems; Human Centered Systems; and Education, Training, and Human Resources. The NCO supports and coordinates the working groups' activities.

Many publications are available on their well-organized Web page for free. These include *Blue Books* and *Implementation Plans* as well as the document, "Information Technology for the Twenty-First Century: A Bold Investment in America's Future."

Open Source Organization
www.opensource.org.

This group has recently formed (1998) to help define the converging resources of the burgeoning open source code movement. As the group defines it, "Open source promotes software reliability and quality by supporting independent peer review and rapid evolution of source code. To be certified as open source, the license of a program must guarantee the right to read, redistribute, modify, and use it freely."

Santa Fe Institute (SFI)
1399 Hyde Park Road
Santa Fe, NM 87501
(505) 984-8800
(505) 982-0565, fax
www.santafe.edu

The Santa Fe Institute is an international research center that was founded in 1984. It brings in scientists from universities and institutions throughout the world. Much of the investigation at this computational center looks at real world problems that may have solutions in complexity theory. Attempts are made, through the development of complex algorithms, to glean underlying patterns in actual physical processes. Computer modeling is a key to this effort.

SeMaTech
SEMATECH
2706 Montopolis Drive
Austin, TX 78741
(512) 356-3086, fax
www.sematech.org/public/home.htm

SEMATECH, which is short for SEmiconductor MAnufacturing TECHnology, is a nonprofit, industry-based, technology development consortium of U.S. semiconductor manufacturers that was founded in 1987. Its mission is to share and support information on evolving techniques and standards of microprocessor manufacturing. Areas of study and research include

- Interconnect
- Front end processes
- Assembly and packaging
- Design systems
- Manufacturing methods
- Lithography infrastructure
- 300 mm wafer tool development
- Standards
- Environment, safety, and health issues
- Manufacturing methods

Companies involved in the consortium include AMD, Compaq, CONEXANT, Hewlett-Packard, IBM, Intel, Lucent Technologies, Motorola, Texas Instruments, SEMATECH members, Hyundai, Philips, STMicroelectronics, Siemens, and TSMC.

Semiconductor Industry Association (SIA)
181 Metro Drive, Suite 450
San Jose, CA 95110
(408)436-6600
(408) 436-6646, fax
http://www.semichips.org/

This industry group was organized to bring together industry manufacturers and suppliers, government organizations, consortia, and universities. The association's main task is to establish the goals of microprocessor development. In support of this effort, the SIA publishes *The Technology Roadmap for Semiconductors* (TRS) on a biannual basis. The publication delineates a common approach that helps to ensure advancements in the performance of integrated circuits.

Society of Women Engineers (SWE)
120 Wall Street, Eleventh Floor
New York, NY 10005-3902
(212) 509-9577
(212) 509-0224, fax
www.swe.org

SWE was founded in 1950 to create an organization that "stimulates women to achieve full potential in careers as engineers and leaders, expands the image of the engineering profession as a positive force in improving the quality of life, and demonstrates the value of diversity." The group boasts a membership of over 15,000 in 85 sections (plus 285 student sections). Key objectives of the SWE, as adopted in 1958 include

- Informing young women, their parents, counselors, and the general public, of the qualifications and achievements of women engineers and the opportunities open to them.
- Assisting women in readying themselves for a return to active work after temporary retirement.
- Serving as a center of information on women in engineering.
- Encouraging women engineers to attain high levels of education and professional achievement.

The organization publishes the *Magazine of the Society of Women Engineers.*

Software Development Forum (SDF)
111 West St. John, Suite 200
San Jose, CA 95113
(408) 494-8378
(408) 494-8383, fax
www.center.org

Located at the center of software development in the United States, Silicon Valley, the SDF is the result of the merger between the Software Entrepreneur's Forum and the Center for Software Development. The SDF is designed to provide developers with information, connection, and education. The organization has more than 1,200 members and support for 20-30 events each month. Strategic support comes from Apple Computer, Inc., Sun Microsystems, the Redevelopment Agency of the City of San Jose, Novell, Inc., and others.

University of Southern California (USC) Information Sciences Institute (ISI)
4676 Admiralty Way, Suite 1001
Marina del Rey, CA 90292-6695
(310) 822-1511
(310) 823-6714, fax
www.isi.edu

ISI was started in April 1972 as an off-campus research facility of USC's School of Engineering. It has evolved into one of the nation's leading university-based information processing research centers. The institute is involved in a wide array of information processing research projects as well as the development of advanced computer and communication technologies and systems. Its director was Internet legend, Jon Postel.

Women in Technology International (WITI)
4641 Burnet Avenue
Sherman Oaks, CA 91403
(818) 990-6705
(818) 906-3299, fax
www.witi.org

Women in Technology International was started in 1989 to increase the number of women in executive positions at technology companies, to help women become more financially independent and technologically literate, and to encourage young women to choose careers in the sciences. It now has a membership of more than 6,000.

World Wide Web Consortium (W3C)
MIT Laboratory for Computer Science
545 Technology Square
Cambridge, MA 02139
(617) 253-5851
(617) 258-8682, fax
www.w3.org

Tim Berners-Lee, creator of the Web, is director of the W3C which was founded in October 1994 "to lead the World Wide Web to its full potential by developing common protocols that promote its evolution and ensure its interoperability." The group is an international industry consortium, now hosted by the Massachusetts Institute of Technology (MIT) Laboratory for Computer Science. The consortium initiatives include acting as a repository of information about the Web for developers and users; promulgating reference code implementations to embody and promote standards; and creating prototype and sample applications to demonstrate use of new technology.

XEROX Palo Alto Research Center (XEROX PARC)
3333 Coyote Hill Road
Palo Alto, CA 94304-1314
(650) 812-4000
www.parc.xerox.com/parc-go.html

In 1970, Xerox Corporation brought together some of the leading information technology engineers and researchers of the day to create "the architecture of information." Over the years, scientists working at PARC did just that. The centers' major accomplishments include

- personal distributed computing
- graphical user interfaces
- the commercial mouse
- bit-mapped displays
- Ethernet
- client/server architecture
- object-oriented programming
- laser printing

CHAPTER TEN
Print and Electronic Resources

T he first section of this chapter lists over 100 books that are useful references on the developments, technicalities, and issues discussed in this volume. An effort has been made to include only works published after 1996 unless an older resource proved especially useful and current. The book section is divided into two subsections. The first part, Computer Science and Technology, lists many works of a technical nature, such as programming guides, telecommunications handbooks, and dictionaries. The second subsection, Computers and Society, includes books that deal more with social computing issues than with the technologies of machines.

Following this section, journals and magazines that document the latest events of the digital age, especially as they relate to computers and telecommunications, are listed. Then, in the final section, a series of useful Web sites that offer the most current information on the subjects are listed. One should be aware that Internet resources often change and disappear quickly. But the targets are included because this is the best source for information on the emerging developments in computers and computer science.

BOOKS

Computer Science and Technology

Booch, Grady, Ivar Jacobson, and James Rumbaugh. *The Unified Modeling Language User Guide*. The Addison-Wesley Object Technology Series. Menlo Park, CA: Addison-Wesley, 1998. 482 pages. ISBN: 0-201-57168-4.

The Unified Modeling Language (UML) standard for documenting software designs is one of the important developments in software engineering in the past few years. Written by UML's developers, *The Unified Modeling Language User Guide* provides a basic tour of the essential concepts and modeling diagrams used in UML. With great detail and excellent advice, this book is essential reading for anyone who wants to know more about UML.

Bowman, Judith S., Sandra L. Emerson, and Marcy Darnovsky. *The Practical SQL Handbook: Using Structured Query Language*. 3rd ed. Book and CD-ROM. Menlo Park, CA: Addison-Wesley, 1996. 454 pages. ISBN: 0-201-44787-8.

This book explains Structured Query Language (SQL) in a simple manner. This is an update to the best-selling second edition. A CD-ROM includes Sybase SQL Anywhere Desktop Runtime, which is an SQL database for the user to achieve some practical experience in SQL programming.

Bradford, Rex E. *Real-Time Animation Tool-Kit in C++*. Book and CD-ROM. New York: John Wiley & Sons, 1995. 778 pages. ISBN: 0-471-12147-9.

Designed for the graphics programmer, this book offers C++ tools and utilities, a C++ animation class library, and useful techniques to teach real-time animation. A CD-ROM is included so the user has access to the class library, important codes for examples in the book, extra programs, and an art gallery.

Burdick, Howard E. *Digital Imaging: Theory and Applications*. Book and CD-ROM. New York: McGraw Hill, 1997. 304 pages. ISBN: 0-079-13059-3.

This book's author, Howard Burdick is recognized as one of the world's leading experts on digital imaging. His guide book supplies hard-to-find information of interest to both graphic artists and programmers. A companion CD-ROM allows the reader a chance to experiment with the techniques the author has explained in print. Readable and concise, this book is a necessary text for those interested in the development of digital art.

Castro, Elizabeth. *HTML 4 for the World Wide Web: Visual QuickStart Guide*. 3rd ed. Berkeley, CA: Peachpit Press, 1997. 336 pages. ISBN: 0-201-69696-7.

This book is a precise documentation of how to write Web pages using HTML (hypertext markup language), version #4. Using full color charts, the author helps the reader see the best way to create common elements of the Web page design including titles, headers, links, adding tables, frames, forms, and multimedia.

Cerami, Ethan. *Delivering Push*. Book and CD-ROM. New York: Computing McGraw-Hill, 1998. 424 pages. ISBN: 0-079-13693-1.

This book and companion CD-ROM include examples and implementations of push technology (delivering scheduled information over the Web). All of the major push servers and clients on the market are included in this book. It covers installation and configuration of BackWeb, Castanet, PointCast, and other Web casting tools.

Cooper, Alan. *About Face: The Essentials of User Interface Design*. Indianapolis, IN: IDG Books Worldwide, 1995. 400 pages. ISBN: 1568843224.

Written by the developer of the Visual Basic programming protocol, this book is recommended for those who want to understand why most software is so poorly designed, or those who want to write better software code. "Must reading (and doing!) for programmers of any level," according to Paul Saffo, director of the Institute for the Future.

Crane, Randy. *A Simplified Approach to Image Processing: Classical and Modern Techniques in C*. Hewlett-Packard Professional Books / Prentice-Hall ECS Professional. Book and Disk. Paramus, NJ: Prentice Hall, 1996. 336 pages. ISBN: 0-13-226416-1.

This book presents a comprehensive overview and introduction to the most popular image processing techniques currently used by professionals. Processing of color images, image warping and morphing techniques, and image compression are thoroughly covered. An accompanying diskette includes real examples for experimenting on the techniques explained in the book.

Dickman, Alan. *Designing Applications with MSMQ: Message Queuing for Developers*. Menlo Park, CA: Addison-Wesley, 1998. 512 pages. ISBN: 0-201-32581-0.

This book is a technical, but readable tome that reveals the "facelift" in the new Microsoft Message Queue Server (MSMQ). This server application, which is part of Microsoft BackOffice for NT 4, lets designers work on flexible communications needs that have become too apparent in today's Internet and distributed systems. The book also includes MSMQ programming samples.

Dodd, Annabel Z. *The Essential Guide to Telecommunications*. Upper Saddle River, NJ: Prentice Hall PTR, 1999. 240 pages. ISBN: 0-13-014295-6.

This book gives a concise explanation of Internet and wireless services including, intranets, extranets, PCS services, wireless data, local telephony, and the new Low Earth Orbit Satellites (LEOS). The book is highly regarded for its clear discussion of the technologies and for its useful coverage of the Telecommunications Act of 1996.

Dougherty, Ray C. *Natural Language Computing: An English Generative Grammar in Prolog*. Hillsdale, NJ: Lawrence Erlbaum, 1994. 349 pages. ISBN: 0-8058-1525-2.

This book is designed for the beginner who wants to learn more about how to use natural language processing, linguistic theory, artificial intelligence, machine translation, and expert systems. The author notes that "the basic idea is to present meaningful answers to significant problems involved in representing human language data on a computing machine. My main focus is on the grammatical devices underlying constructions of English, French, and German."

D'Souza, Desmond F. and Alan Cameron Wills. *Objects, Components and Frameworks with UML: The Catalysis Approach*. Addison-Wesley Object Technology Series. Berkeley, CA: Peachpit Press, 1998. 816 pages. ISBN: 0-201-31012-0.

Unified Modeling Language (UML) is the focus of this book, which provides expert advice on creating better design documents and components.

The author relies on the Catalysis design protocol (a new method for constructing open component systems from frameworks) to offer some advantages in designing for recent software requirements.

Dubois, Didier, Henri Prade, and Ronald R. Yager, eds. *Fuzzy Information Engineering: A Guided Tour of Applications*. New York: John Wiley & Sons, 1996. 712 pages. ISBN: 0-471-14766-4.

Fuzzy logic, the tool that allows computer programmers to understand nebulous commands that ordinary programs are unable to decipher, is the subject of this book. With the implementation of fuzzy logic, computers can control reactionary objects like automobiles. Each chapter explores fuzzy logic using real-life applications.

Emmerson, Bob and David Greetham. *Computer Telephony and Wireless Technologies: Future Directions in Communications*. Charleston, SC: Computer Technology Research Corp., 1997. 275 pages. ISBN: 1-56607-992-6.

Computer telephony (CT) and wireless communications are two of the most important new developments in digital connectivity. This Computer Technology Research Corporation (CTR) report focuses on the business applications of CT and wireless communications and explains their implementation and benefits. The book defines CT "as a platform that merges voice and data services in order to enable the development of integrated business applications."

Freedman, Alan. *The Computer Glossary: The Complete Illustrated Dictionary*. Book and CD-ROM. New York: AMACOM, 1998. 700 pages. ISBN: 0-8144-7978-2.

This is one of the many dictionaries available that explain the various terms and acronyms associated with computers and digital communications. This book also deals with some of the slang generated by cyber society and online communities. A companion, searchable CD-ROM is included.

Garcia, Narciso, Arthur Damask, and Steven Schwarz. *Physics for Computer Science Students: With Emphasis on Atomic and Semiconductor Physics*. New York: Springer-Verlag, 1998. 550 pages. ISBN: 0387949038.

Grady, Sean M. *Virtual Reality: Computers Mimic the Physical World*. Springer-Verlag Science Sourcebook series. New York: Facts on File, Inc., 1998. 144 pages. ISBN: 0816036055.

As a part of Springer-Verlag's Science Sourcebook series, this book provides a clear explanation of virtual reality (VR). The work includes a presentation of VR history and possibilities for the future. Little is said about the possible psychological effects associated with living in a virtual world, but the state of VR is well documented.

Grand, Mark. *Patterns in Java*. Vol. 1. Book and CD-ROM. New York: John Wiley & Sons, 1998. 640 pages. ISBN: 0-471-25839-3.

This book includes a library of 41 software design patterns for class design in Java. The process is illustrated using Unified Modeling Language (UML) techniques and includes sample Java code. The author has compiled the best Java patterns in the past five years and made them available in a user-friendly publication.

Gregory, Donald and J. Regis Bates. *Voice and Data Communications Handbook: Signature Edition*. New York: McGraw Hill, 1998. ISBN: 0-070-06396-6.

According to one reviewer, this book "is the most comprehensive, up-to-date and jargon free reference in a rapidly changing field." Since its subject is one of the world's fastest-growing industries, accurate information for a broad cross section of users and potential employees is critical. Areas of inquiry include voice communications, telephone equipment and networks, analog vs. digital transmission, lines vs. trunks, service carriers, and traffic engineering.

Grimes, Richard. *Professional ATL COM Programming*. Chicago, IL: Wrox Press, 1998. 500 pages. ISBN: 1-86100-140-1.

This book is written by an experienced Active Template Library (ATL) developer and is geared to advanced C++ programmers.

Grimes, Richard, Alex Stockton, George Reilly, and Julian Templeman. *Professional ATL COM Programming*. Chicago, IL: Wrox Press, 1998. 703 pages.

Active Template Library (ATL) is the part of the Microsoft Foundation Classes (MFC) used to build reusable and efficient components in the C++ computer language. While use of ATL COM can be a challenge, this book is designed to give C++/MFC programmers assistance in this useful area.

Gruber, Martin. *Understanding SQL*. Alameda, CA: Sybex, 1990. 434 pages. ISBN: 0895886448.

This book includes an introduction to the world of relational databases that then lays out a roadmap to the use of SQL (structured query language), which is an industry-standard language for creating, updating, and querying relational database management systems. Chapters are sequential with an emphasis on SQL learning procedures and routines.

Harel, David. *Algorithmics: The Spirit of Computing*. 2nd ed. Menlo Park, CA: Addison-Wesley, 1992. 476 pages. ISBN: 0-201-50401-4.

This book explains the basic ideas of algorithms, their structures, and how they are utilized by computer programs to change data. It is written for an audience of programmers, systems analysts, system designers, and software engineers, but at a level accessible enough to be useful for readers with little mathematics or computer experience.

Harold, Elliotte Rusty. *XML: Extensible Markup Language*. Book and CD-ROM. Indianapolis, IN: IDG Books Worldwide, 1998. 426 pages. ISBN: 0764531999.

Explaining the Extensible Markup Language (XML) system in a straightforward linear manner, this is a useful text for Web page designers hoping to use the new XML tools quickly for dynamic results.

Harris, Stuart and Gayle Kidder. *Official Netscape Dynamic HTML Developer's Guide: Windows & Macintosh*. Book and CD-ROM. Research Triangle Park, NC: Ventana Communications Group, 1997. 344 pages. ISBN: 1566047978.

The companion CD-ROM includes scripts, examples, shareware programs, HTML templates, backgrounds, and other programs that help the user implement some of the procedures learned by reading the accompanying book on the implementation of DHTML protocols.

Haykin, Simon. *Communication Systems*. New York: John Wiley & Sons, 1994. 896 pages. ISBN: 0-471-57176-8.

The reader of this book should have a knowledge of electronics, circuit theory, and probability theory. This knowledge, of course, is a helpful prerequisite for anyone choosing to study communications systems. This book is an accepted text for a course that surveys analog and digital communications systems.

Hernandez, Michael J. *Database Design for Mere Mortals: A Hands-On Guide to Relational Database Design*. Menlo Park, CA: Addison-Wesley. 480 pages. ISBN: 0-201-69471-9.

This book is a practical guide that includes both the theoretical philosophy behind the design of relational databases and the real world application of effective methodology. The author is an experienced teacher accustomed to providing hands-on experience for his students.

Hutchinson, Sarah E., Stacey C. Sawyer, and Glen J. Coulthard. *Computers, Communications, and Information: A User's Introduction*. Rev. ed. Core Version. New York: Irwin/McGraw-Hill, 1998. ISBN: 0-071-09327-3.

Recommended as an introductory text for information systems teachers and students, this book provides exceptional information in a readable, easily understood format.

Illingworth, Valerie, ed. *Dictionary of Computing*. 4th ed. New York: Oxford University Press, 1997. 576 pages. ISBN: 0-19-853855-3.

Designed as a useful reference tool for computer users, students, teachers, and computer industry professionals alike, this dictionary includes most computer, networking, communications, and Internet terms. Almost 4,500 listings are included in this edition, which has been expanded considerably since the first edition appeared in 1983.

Irwin, James H. and James Harry Green. *The Irwin Handbook of Telecommunications*. Burr Ridge, IL: Irwin Professional Publishing, 1996. 1,051 pages. ISBN: 0786304790.

This book is an excellent, concise reference guide to North American telecommunications systems. It includes sections on broadband distribution systems and Internet telephony as well as comprehensive overviews of common and emerging communications standards.

Keyes, Jessica. *Datacasting: How to Stream Databases over the Internet*. New York: Computing McGraw-Hill, 1998. 512 pages. ISBN: 0-070-34678-X.

The interface between databases and HTML has become an important mechanism for presenting Web-based information. The author explains how to use common database software from Informix, Oracle, Sybase, IBM, Microsoft, and even shareware developers to create dynamic, streaming presentations.

Liberty, Jesse. *Teach Yourself C++ in 21 Days*. 2nd ed. Indianapolis, IN: Sams Publishing, 1997. 792 pages. ISBN: 0672310708.

Through the use of tutorials and reasonably paced lessons, this classic guide to computer programming makes the learning of C++ almost simple. The readers will stay motivated as they see fast results. The book requires no previous programming expertise to use it.

Loney, Kevin and George B. Koch. *Oracle 8: The Complete Reference*. Book and CD-ROM. New York: Oracle Press, 1997. 1,300 pages. ISBN: 007882396X.

This book is an updated text that covers versions 7.0 through 8 of Oracle, the popular database application. It is useful for those looking for detailed reference data on how to set up and maintain an Oracle database. Look for the next iteration which should include Oracle 8*i*, the Internet centric configuration.

Maguire, Steve. *Debugging the Development Process: Practical Strategies for Staying Focused, Hitting Ship Dates, and Building Solid Teams*. Redmond, WA: Microsoft Press, 1994. 183 pages. ISBN: 1-5561-5650-2.

Anyone involved in programming can use the practical advice contained in this book by one of software development's experts. The author uses his experience as an organizer and manager of software development teams at Microsoft to explain how to bring about timely applications.

Margolis, Philip E. *Random House Personal Computer Dictionary*. 2nd ed. New York: Random House, 1996. 528 pages. ISBN: 0-679-76424-0.

This is another computer user dictionary that keys on practical terms (approximately 1,900 entries in this edition) to help even novices get through the maze of computer-related words that have been developed in recent years.

McCarthy, Jim, Steve McConnell, and Steve Maguire. *Software Engineering Classics*. Redmond, WA: Microsoft Press, 1998. 712 pages. ISBN: 0-7356-0597-1.

This three-book set includes Steve Maguire's *Debugging the Development Process* (see above), Jim McCarthy's *Dynamics of Software Development*, and Steve McConnell's *Software Project Survival Guide*. Together, these three texts are essential for anyone involved in the software development process.

McConnell, Steve. *Code Complete: A Practical Handbook of Software Construction*. Redmond, WA: Microsoft Press, 1993. 857 pages. ISBN: 1-5561-5484-4.

This book has been recommended by a reviewer as the "best practical guide to writing commercial software," and is highly recommended for anyone interested in learning the process. The intent is to have the reader write clearer, more efficient code in less time. As McConnel says in the preface, "my primary concern in writing this book has been to narrow the gap between the knowledge of industry gurus and professors on one hand and common commercial practice on the other. Although leading-edge software-development practice has advanced rapidly in recent years, common practice hasn't."

Microsoft Corporation. *Microsoft Press Computer Dictionary*. 3rd ed. Book and CD-ROM. Redmond, WA: Microsoft Press, 1997. 704 pages. ISBN: 1-5723-1446-X.
Especially useful because of the companion, fully searchable CD-ROM, *Microsoft Press Computer Dictionary* defines terms pertaining to all areas of computing. It can be used by both beginning and advanced computer users.

Microsoft Corporation. *Microsoft Visual C++ 6.0 Reference Library*. Redmond, WA: Microsoft Press, 1998. 5,072 pages. ISBN: 1-5723-1865-1.
If one wishes to program using the C++ language, this is an absolutely critical resource. The multivolume set is a comprehensive resource from the Microsoft Development team for programming in the object-oriented development environment.

Miles, Peggy. *Internet World Guide to Webcasting*. New York: John Wiley & Sons, 1998. 416 pages. ISBN: 0-471-24217-9.
Webcasting or sending television or radio content over the Internet or a local area network is introduced in this book. Written by an expert in this developing field, this book provides many examples of how corporations and businesses are using the technology today for training, promotion, etc.

Mudry, Robert J. *The DHTML Companion*. Paramus, NJ: Prentice Hall Computer Books, 1998. 400 pages. ISBN: 0-13-796046-8.
This book covers the new Web standard known as Dynamic HMTL (DHTML), which makes it easier to add multimedia content to Web pages. Cascading style sheets that control how fonts are displayed, an improved scripting language, and Document Object Model (DOM) that allows access to all elements are explained in this book.

Mueller, John Paul and Anthony Gatlin. *The Complete Microsoft Certification Success Guide*. 2nd ed. Book and CD-ROM. New York: Computing McGraw-Hill, 1997. 384 pages. ISBN: 0-079-13201-4.
Coauthored by a Microsoft Corporation professional, this guide to all of Microsoft's four certification programs has been updated recently and includes an interactive CD-ROM with assessment tests for many of the exams. It includes tips on taking the exams and advice on how to study and practice. Since Microsoft certification is growing at the rate of 350%, this book provides a necessary resource to those interested in becoming certified.

Muller, Nathan J. *Desktop Encyclopedia of Telecommunications*. New York: Computing McGraw-Hill, 1998. 550 pages. ISBN: 0-070-44457-9.
Since it concentrates on the complex and rapidly-evolving world of telecommunications, this encyclopedia is an indispensable asset to students or professionals in the field.

Negrino, Tom and Dori Smith. *JavaScript for the World Wide Web: Visual QuickStart Guide*. 3rd ed. New York: Peachpit Press, 1999. 292 pages. ISBN: 0-201-35463-2.

Javascript programming is the next step for those who got their first experience in programming by writing HTML files for display on Web browsers. This book is a good introductory text for anyone who would like to learn the basics of Javascript.

Netravali, Arun N. and Barry G. Haskell. *Digital Pictures: Representation, Compression, and Standards.* 2nd ed. New York: Plenum, 1995. 706 pages. ISBN: 030644917X.

The Wall Street Journal has called this book "today's bible on video compression." This updated edition covers fundamentals, algorithms, and standards for digital imagery, including the emerging standards for high definition television (HDTV) and multimedia PCs. *Digital Pictures* looks at formats such as JBIG, JPEG, H.261, CCIR601, CCIR723, MPEG1, and MPEG2.

Newton, Harry. *Newton's Telecom Dictionary.* 15th ed. San Francisco: Miller Freeman, 1999. 901 pages. ISBN: 1578200318

This dictionary is probably the most respected and used telecommunications reference. The author is so intent on keeping his dictionary current that he adds nearly 100 terms each week and issues updates every six months.

O'Leary, Timothy J. and Linda I. O'Leary, *Computing Essentials: Multimedia Edition 1997-1998.* 1999–2000 ed. Book and CD-ROM. New York: McGraw Hill, 2000. ISBN: 0-073-65556-2.

The authors have written an interactive text with companion CD-ROM that will teach the reader the basics of computer use. This book explores rudimentary and advanced application software, the operating system, microprocessors, input and output, storage systems, and communications.

Pfaffenberger, Bryan. *Webster's New World Dictionary of Computer Terms.* 6th ed. Indianapolis, IN: Macmillan General Reference, 1997. 592 pages. ISBN: 0028618904.

The list of terms in this edition of the dictionary has reached over 3,500 with extensive reference to Internet and cyber-societal slang. The book also includes some 100 illustrations that help to clarify concepts.

Purcell, Lee. *Internet Audio Sourcebook.* Book and CD-ROM. New York: John Wiley & Sons, 1997. 553 pages. ISBN: 0-471-19150-7

This book is a good overall introduction to the various utilities, applications, and standards used to stream audio data across the Internet. MIDI, WAV, AU, and other audiobinary protocols are thoroughly covered. The book includes a CD-ROM with demo software, scripts, and various useful code for experimentation.

Sampei, Seiichi. *Applications of Digital Wireless Technologies to Global Wireless Communications.* Paramus, NJ: Prentice Hall, 1997. 350 pages. ISBN: 0-13-214272-4.

The author is considered the father of Japanese wireless technology. His work in this book provides a complete overview of the next generation of technology (including digital multimedia over wireless handsets). Designed for communications professionals, the book presents many real-life applications and scenarios to teach this subject.

Shapiro, Jeffrey R. *Computer Telephony Strategies.* Indianapolis, IN: IDG Books Worldwide, 1996. 420 pages. ISBN: 0764530135.

The author is a telephony expert who has written about strategies used to integrate voice over the network. A reader reports, "the anecdotes, tone, and structure of this book make it both an extremely pleasurable experience as well as an invaluable learning tool."

Sirockman, Jason. *MCSE Complete.* 2 Manuals, 2 CDs. Indianapolis, IN: New Riders, 1998. 900 pages. ISBN: 0789715643.

Microsoft has authorized this complete interactive package (2 manuals, 2 audio CDs, and examination questions and answers) for training in the company's certification program for professionals. Courses included are Windows NT Server 4.0, Win NT Server in the Enterprise 70-58 Networking Essentials, Windows NT Workstation, Windows 95, Microsoft TCP/IP, Exchange Server 5.0, Microsoft SMS 1.2, Microsoft SQL Server 6.5 Admin, and Microsoft SQL 6.5 Server implementation.

Stone, John A. *Developing Software Applications in a Changing IT Environment: Management Strategies and Techniques.* New York: McGraw Hill, 1997. 320 pages. ISBN: 0-070-61719-8.

Because the technologies and end-use applications are changing so rapidly, the methods being employed to create useful and efficient software are also multiplying at an accelerated rate. The author delineates some methods that developers can use to put a semblance of order into the process.

Taylor, Ed. *TCP/IP Complete.* Book and CD-ROM. New York: Computing McGraw-Hill, 1998. 800 pages. ISBN: 0-070-63400-9.

The author of this book is a network architect and consultant who uses this book as a platform to explain TCP/IP systems, components, and applications in a very succinct tone. He uses many illustrations and charts to help the potential network designer stay focused on the development tasks at hand.

Teague, Jason C. *DHTML: For the World Wide Web.* Visual Quickstart Guide Series. Menlo Park, CA: Addison-Wesley, 1998. 248 pages. ISBN: 0-201-35341-5.

Another one of the popular Visual Quickstart Guide series, this text covers the most recent code and structures for Dynamic HTML (DHTML) developers in a simple and straightforward style. While programmers will find the book useful, it is directed more at Web page designers who want to include more multimedia content on their sites.

Tomasi, Wayne. *Advanced Electronic Communications Systems.* 4th ed. Paramus, NJ: Prentice Hall, 1998. 413 pages. ISBN: 0-13-750986-3.

The three-book set included here covers the whole range of communications technologies from electronic communications fundamentals to the modern digital and data communications systems. The author suggests that knowledge and experience with basic electronic principles and mathematics are key prerequisites for understanding its concepts.

Walters, Robert. *Computer Telephony Integration.* 2nd ed. Norwood, MA: Artech House, 1998. 412 pages. ISBN: 0890069697.

If the reader is in need of a "well-informed analysis of where computer telephony integration (CTI) came from and where it is going," this update of the classic first edition is an excellent resource.

Computers and Society

Agre, Philip and Marc Rotenberg, eds. *Technology and Privacy: The New Landscape.* Cambridge, MA: MIT Press, 1997. 280 pages. ISBN: 0-262-01162-X.

This compilation contains 10 scholarly essays on the current state of technology-based privacy issues. Written from the points of view of communications, human-computer interaction, law, political science, and sociology, each contributor looks to concerns that will affect the future. Emphasis is placed on cybertechnologies which can be either threats to or protection for personal privacy in the Information Age.

Alliance for Technology Access Staff. *Computer Resources for People with Disabilities: A Guide to Exploring Today's Assistive Technologies.* 3rd rev. "Millennium ed." Alameda, CA: Hunter House, 2000. ISBN: 089793301.

Altman, Don B. and Steven Ray (Illustrator). *Digital Frontier Job & Opportunity Finder: Tomorrow's Opportunities Today.* Redondo Beach, CA: Moon Lake Media, 1996. ISBN: 0963916114.

Brockman, John. *Digerati: Encounters with the Cyber Elite.* San Francisco: Hardwired, 1996. 354 pages. ISBN: 1888869046.

The author of this book spent August 1995 through April 1996 recording conversations with 33 of the most important developers of cyberspace. He condenses their conversations in *Digerati*, providing insight into the lives of some of the most influential digital visionaries of the last quarter century.

Cairncross, Frances. *The Death of Distance: How the Communications Revolution Will Change Our Lives.* Cambridge, MA: Harvard Business School Press, 1997. 303 pages. ISBN: 0875848060.

Written by a well-respected economist, this book looks at the changes technology has wrought on our economic institutions and conventions. The author posits that the ability to communicate online means that distance has no meaning (and is no longer a barrier) where most considerations of business are concerned.

Cate, Fred H. *Privacy in the Information Age.* Washington, DC: Brookings Institution Press, 1997. 200 pages. ISBN: 0815713150.

The author, a professor at Indiana University School of Law-Bloomington, wrote this book as an analysis of the privacy issues that digital advances have introduced into society. This book explains how the fields of business, communications, economics, and law are affected by the abundance of data available via the Internet and how this impacts all of society in the United States. The book concludes that there is little consensus on how to deal with the concerns raised.

Cavoukian, Ann and Don Tapscott. *Who Knows: Safeguarding Your Privacy in a Networked World.* New York: McGraw-Hill, 1996. 233 pages. ISBN: 0-070-63320-7.

Written by two experts in the field of cyberspace privacy, this analysis of the methods being employed to gather personal information about you and your family on the Internet is full of frightening examples. To avoid just a gloom and doom scenario, the authors also include methods that can be employed to guard against the loss of privacy.

Computers : Career in Focus. Careers in Focus Series. Chicago, IL: Ferguson, 1998. ISBN: 0894342207.

Christianson, J. Scott and Ava Fajen. *Computer and Network Professional's Certification Guide.* Alameda, CA: Sybex, 1996. 560 pages. ISBN: 0782122604.

This book offers an overview of approximately 70 certification programs now available for professionals in the computer and information technology industry.

Crevier, Daniel. *AI: The Tumultuous History of the Search for Artificial Intelligence.* New York: Basic Books, 1994. 386 pages. ISBN: 0465001041.

AI (artificial intelligence) has had a less than glorious history since the first experimentation in the early 1950s. This book provides a sufficient basic history of the developments in this rarified world of computer science, and includes many true stories from actual participants in the progress of AI over the last 50 years.

De Jager, Peter and Richard Bergeon. *Managing '00: Surviving the Year 2000 Computing Crisis.* New York: John Wiley & Sons, 1997. 224 pages. ISBN: 0-471-17937-X.

By describing available tools and recommending how a business might best plan to react to any contingencies that arise from the dreaded millennium bug, *Managing '00* is one of better resources for those researching the Y2K problem. The authors are considered experts in the field, with De Jager being credited as one of the first professionals in the field to raise the warning flag in 1993 about the potential Y2K crisis.

Denning, Peter and Robert Metcalfe, eds. *Beyond Calculation: The Next Fifty Years of Computing.* New York: Copernicus, 1997. 313 pages. ISBN: 0-387-98588-3.

The editors, Bob Metcalfe, the inventor of Ethernet technology and founder of the 3Com corporation and Peter J. Denning, chair of the Computer Science Department in the School of Information Technology and Engineering at George Mason University, present a series of essays on the remarkable and far-reaching impacts of digital machines on our society.

Dery, Mark. *Escape Velocity: Cyberculture at the End of the Century.* Boston: Grove/Atlantic, 1997. 400 pages. ISBN: 080213520X.

High tech cyberpunks, cyberhippies, cybersexers, technopagans, and cyborgs are a few of the subcultures of the online world explored in this odd review of the possible technological future. The author makes the argument that cyberculture is reaching an escape velocity that will free it from the constraints of history and accepted cultural conventions.

DiBona, Chris, Sam Ockman, and Mark Stone, eds. *Open Sources: Voices from the Open Source Revolution*. Sebastopol, CA: O'Reilly & Associates, 1999. 280 pages. ISBN: 1-56592-582-3.

For this book, leaders in the Open Source movement contributed essays on the history, philosophy, and future of open source. The book explains why free code runs the majority of Internet servers and how some key applications have surpassed the best efforts of giant software distributors like Microsoft. The contributors include a who's who of freeware, such as Brian Behlendorf (Apache), Kirk McKusick (Berkeley Unix), Tim O'Reilly (Publisher, O'Reilly & Associates), Bruce Perens (Debian Project, Open Source Initiative), Tom Paquin and Jim Hamerly (mozilla.org, Netscape), Eric Raymond (Open Source Initiative), Richard Stallman (GNU, Free Software Foundation, Emacs), Michael Tiemann (Cygnus Solutions), Linus Torvalds (Linux), Paul Vixie (Bind), Larry Wall (Perl), and Bob Young (Red Hat).

Diffie, Whitfield and Susan Eva Landau. *Privacy on the Line: The Politics of Wiretapping and Encryption*. Cambridge, MA: MIT Press, 1998. 352 pages. ISBN: 0-262-04167-7.

Written to point out that "the preservation of privacy is critical to a democratic political process," *Privacy on the Line* argues that this goal is becoming more difficult as we spend increasing hours of the day using telephones, computer networks, and making electronic transactions of all kinds. The authors look at ways of preserving privacy and predict negative consequences if care is not taken to protect citizens' rights to a secure communications environment. Of main interest is the need for computer encryption, of which Diffie is considered a national expert.

Dyson, Esther. *Release 2.0: A Design for Living in the Digital Age*. New York: Broadway Books, 1997. 224 pages. ISBN: 0767900111.

This is the first book by one of the most respected thinkers of the digital domain. Dyson, CEO of Edventure Holdings also publishes *Release 1.0*, an industry newsletter. *Release 2.0* deals with computing and the Internet and how advances in their development have changed and will continue to change our social, economic, and political interactions. *Release 2.1: A Design for Living in the Digital Age (An Upgrade for Living in the Digital Age)* is the follow up to this book, published less than one year afterward.

Eberts, Marjorie and Margaret Gisler. *Careers for Computer Buffs & Other Technological Types*. Vgm Careers for You Series. Lincolnwood, IL: Vgm Career Horizons, 1998. ISBN: 0844247073.

Edstrom, Jennifer and Marlin Eller. *Barbarians Led by Bill Gates: Microsoft from the Inside*. New York: Henry Holt & Company, 1998. 304 pages. ISBN: 0805057544.

This very popular book is written by two former Microsoft insiders who question the company's business practices and ethics. Although critical of Microsoft, the authors still provide a thorough history of developments at Microsoft from the 1980s to the present day by concentrating on the key technologies and personalities involved.

Epstein, Richard G. *The Case of the Killer Robot: Stories about the Professional, Ethical, and Societal Dimensions of Computing.* New York: John Wiley & Sons, 1996. 242 pages. ISBN: 0-471-13823-1.

Focusing on the ethical issues in the computer science profession, this book is a rather unusual work. It is presented as a fictional story with the central character being a robot that kills a human worker after it has been programmed to automate an assembly. The author ponders the question what is the human factor?

Gates, Bill, Nathan Myhrvold, and Peter M. Rinearson. *The Road Ahead.* Book and CD-ROM. New York: Penguin, 1996. 332 pages. ISBN: 0140260404.

This book is an excellent mapping of Bill Gates's vision for computing's future. Written with the help of his chief technology aide Myhrvold, the book makes some important predictions and includes some important points about the development of Internet computing.

Gelernter, David Hillel. *Machine Beauty: Elegance and the Heart of Technology.* New York: Basic Books, 1998. 160 pages. ISBN: 0465045162.

The author of this book is also known for being a target of the Unabomber and was seriously injured in one of his mail-bomb attacks in 1993. In *Machine Beauty*, he attempts to reconcile the world of art and beauty with that of computer designers and programmers and their allegedly poorly designed computer devices. The author contends that programmers have had inadequate training in the disciplines that would help them create more elegant and useful human-computer interfaces.

Goldberg, Jan. *Great Jobs for Computer Science Majors.* Lincolnwood, IL: NTC/ Contemporary, 1997. 240 pages. ISBN: 0844247464.

Goldberg presents a thorough overview of the types of employment one might expect to attain after receiving training in computer science or information technology.

Hobart, Michael E. and Zachary Sayre Schiffman. *Information Ages: Literacy, Numeracy, and the Computer Revolution.* Baltimore, MD: Johns Hopkins University Press, 1998. 328 pages. ISBN: 080185881X.

While they believe that the computer revolution is significantly impacting the world, the authors of this book believe that this "third information age" will be less important to the development of culture than the first two. The first age, they posit, began with the invention of writing and the second with the invention of the printing press.

Jenkins, Timothy L., Andrew Young, and Khafra K. Omraseu. *Black Futurists in the Information Age: Vision of a 21st Century Technological Renaissance.* San Francisco: KMT Publications, 1997. 284 pages. ISBN: 0963564560.

The authors predict that the fast pace of the changes brought about by the current information revolution will provide black Americans with the means to achieve some level of equality in society. . While they see that the current situationlimits minority group's access to opportunity and acceptance, they also see this as a time of untold opportunities for minorities.

Johnson, Steven A. *Interface Culture: How New Technology Transforms the Way We Create and Communicate.* New York: HarperCollins, 1997. 272 pages. ISBN: 0062514822

Kahin, Brian and Charles Nesson, eds. *Borders in Cyberspace: Information Policy and the Global Information Infrastructure.* Publication of the Harvard Information Infrastructure Project. Cambridge, MA: MIT Press, 1997. 300 pages. ISBN: 0-262-11220-5.

This book is a collection of essays that delve into the thorny problems inherent in the application of cyberspace, a medium that defies national boundaries. Whose laws apply when legislation meant for old types of communication are no longer appropriate? Intellectual property, commerce, security, privacy, and censorship are among the issues looked at in this work.

Kaku, Michio. *Visions: How Science Will Revolutionize the 21st Century.* New York: Doubleday, 1998. 416 pages. ISBN: 0-385-48499-2.

The author offers a glimpse into the exciting research that will someday change life on Earth as we know it today. Concentrating on the three scientific fields that reshaped the twentieth century—the quantum, biogenetic, and computer revolutions—the author interviews 150 leading scientists to get their impressions of what will happen in the twenty-first century.

Kaufmann III, William J. and Larry L Smarr. *Supercomputing & the Transformation of Science.* New York: W.H. Freeman & Company, 1993. 238 pages. ISBN: 0716750384.

While some of the material in this book is dated, it is an important and useful work coauthored by Smarr who is the current head of the National Computational Science Alliance, the association of supercomputing research centers, and corporate partners charged with developing the Technology Network Grid.

Kroker, Arthur and Marilouise Kroker, eds. *Digital Delirium.* New York: St. Martin's Press, 1997. 324 pages. ISBN: 0312172370.

This book brings together some of the most interesting and respected experts of digital-based culture to report on their interpretations of what is being wrought by hyper connectivity and cyber communications. While the book is partly an attack on the conservative libertarian use of cyberspace, it goes beyond this to show how the subjects lives are altered by the online experience.

Loader, Brian, ed. *The Governance of Cyberspace: Politics, Technology and Global Restructuring.* New York: Routledge, 1997. 264 pages. ISBN: 0-415-14724-7.

Contributors to this book include political theorists, computer scientists, social theorists, science fiction writers, psychologists, and sociologists who try to outline the issues involved in placing some sort of control on a medium of social exchange that was designed to be technically anarchistic. Issues explored include privacy versus protection, First Amendment rights versus hate groups on the Internet, and other privacy issues.

Ludlow, Peter and Mike Godwin, eds. *High Noon on the Electronic Frontier: Conceptual Issues in Cyberspace*. Cambridge, MA: MIT Press, 1996. 536 pages. ISBN: 0-262-62103-7.

This collection of 33 essays deals with the ongoing issues that have developed as humans claim their acreage in cyberspace. The author offers many useful sources from newsgroup discussion archives to scholarly texts in an effort to present a balanced interpretation of the action online.

Martinez, Anne. *Get Certified and Get Ahead*. New York: Computing McGraw-Hill, 1998. 488 pages. ISBN: 0-070-41127-1.

Though there are many books that explain how to take the tests and prepare for passing the certification process, *Get Certified and Get Ahead* explains what the certification means in the first place. The author covers the basics of the Microsoft Certified Systems Engineer (MCSE), the Microsoft Certified Solution Developer (MCSD), and the Certified NetWare Engineer (CNE) certification process.

Norman, Donald A. *The Invisible Computer: Why Good Products Can Fail, the Personal Computer Is So Complex, and Information Appliances Are the Solution*. Cambridge, MA: MIT Press, 1998. 340 pages. ISBN: 0-262-14065-9.

According to the author of this analysis of the problems of PC design, "the personal computer is perhaps the most frustrating technology ever....It should be quiet, invisible, unobtrusive." His solution to the interface problem lies in carefully crafted devices designed to do one or two specific jobs that are interconnected with other such appliances.

Oakes, Elizabeth H. *Career Exploration on the Internet: A Student's Guide to More Than 300 Web Sites!* Student ed. Chicago, IL: Ferguson Publishing, 1998. 208 pages. ISBN: 0894342401.

While not strictly a resource for finding technology and computer-related jobs on the Internet, this book will prove helpful to students looking for those types of careers.

Perelman, Michael. *Class Warfare in the Information Age*. New York: St. Martin's Press, 1998. 160 pages. ISBN: 0312177585.

Class conflict is not dead according to the author, who tries to debunk the populist view that the Information Age will bring equality to the masses when everyone is connected via computers and the Internet. He warns that as the power of corporations grows, the rights and privileges of individuals will recede.

Peterson's Guide to Colleges for Careers in Computing. Book and 2 disks. Princeton, NJ: Peterson's Guides, 1996. ISBN: 1560795271.

One of the books in the popular Peterson's Guide series, this version is a resource for those looking at the requirements for attendance at schools noted for their computer science programs.

Planning for National Research Council. *Bridge Builders: African Experiences with Information and Communication Technology*. Washington, DC: National Academy Press, 1996. 290 pages. ISBN: 0309054834.

This technical analysis presents reports on how various technologies have been successfully introduced in the scientific and engineering communities of African states.

Plunkett, John and Louis Rossetto, eds. *Mind Grenades: Manifestos from the Future*. San Francisco: Wired Books, 1996. Unpaged. ISBN: 1-888869-00-3.

This fully illustrated book by the editors of *Wired* magazine is a light-hearted, but well-researched look at the possibilities technology will introduce in the future.

Porter, David, ed. *Internet Culture*. New York: Routledge, 1997. 279 pages. ISBN: 0-415-91684-4.

The editor of this book has gathered the works of 15 well-respected scholars to report on new cultures emerging on the Internet. Virtual communities, virtual bodies, the use of language in cyberspace, and Internet politics are the focus of the included writings.

Schneier, Bruce and David Banisar, eds. *The Electronic Privacy Papers: Documents on the Battle for Privacy in the Age of Surveillance*. New York: John Wiley & Sons, 1997. 747 pages. ISBN: 0-471-12297-1.

Bruce Schneier, a security expert, and David Banisar, a privacy advocate, have presented a collection of known, as well as previously secret, government and industry documents that provide telling details of the battle between privacy and surveillance of personal information.

Spinello, Richard A. *Case Studies in Information and Computer Ethics*. Paramus, NJ: Prentice Hall, 1997. 285 pages. ISBN: 0-13-533845-X.

This book is a collection of 38 case studies involving information issues and computer ethics questions. Readers with a great deal of technical expertise will feel comfortable reading this offering and novice users will be able to glean much from its contributors.

Stone, Allucquere Rosanne. *The War of Desire and Technology at the Close of the Mechanical Age*. Cambridge, MA: MIT Press, 1996. 212 pages. ISBN: 0-262-69189-2.

The author is the director of the Interactive Multimedia Lab at the University of Texas, Austin. In this book, he explores the new ways communications technologies (e-mail, Internet chatrooms, etc.) are putting pressure on the traditional sense of personal identity, especially as this relates to gender and sexual roles.

Ulrich, William M. and Ian S. Hayes. *The Year 2000 Software Systems Crisis: Challenge of the Century*. Paramus, NJ: Prentice Hall Computer Books, 1997. 340 pages. ISBN: 0-13-655664-7.

The sensational title is designed to sell books, while the authors' approach is designed to provide practical analysis of an important problem facing programmers and managers. The book's focus is on determining how to control the changeover process that the Y2K bug necessitates.

Wallace, James. *Overdrive: Bill Gates and the Race to Control Cyberspace*. New York: John Wiley & Sons, 1998. 320 pages. ISBN: 0-471-29106-4.

James Wallace is a popular reporter for the Seattle *Post-Intelligencer* and author of *Hard Drive: Bill Gates and the Making of the Microsoft Empire*. (See entry below.) *Overdrive* is a continuation of the saga of Bill Gates as he shifts gears to take advantage of the Internet. The author points out that the emergence of the "network of networks" almost caught Microsoft completely unaware.

Wallace, James and Jim Erickson. *Hard Drive: Bill Gates and the Making of the Microsoft Empire*. New York: HarperBusiness, 1993. 448 pages. ISBN: 0887306292.

The authors present an unvarnished picture of the world's richest man as he directs Microsoft in an attempt to dominate the information technology marketplace. While often unflattering to Gates, the report is a good documentation of the successes as well as the failures of ITs most important company.

Whittle, David B. *Cyberspace: The Human Dimension*. New York: W. H. Freeman & Co., 1997. 456 pages. ISBN: 0716783118.

The author addresses the problems and potential of cyberspace in this engrossing and thorough exploration. He first offers an overview of what cyberspace is and how its norms and customs have developed. He then moves on to discuss how participation in cyberspace affects the individual, community, business, and government in all of society. His own beliefs are interjected at times, and unfortunately he can't resist some not-so-subtle pushing of his views in an otherwise first-rate discussion of vice on the Net. For the most part, Whittle performs a valuable and rare service by resisting easy answers in favor of judicious recommendations.

Wieners, Brad and David Pescovitz, eds. *Reality Check*. San Francisco: Hardwired, 1996. 161 pages. ISBN: 1888869038.

Reality Check is based on the popular futurism section of *Wired*, the magazine of all things digital. It presents the interpretations of 150 experts as they try to decide what the future will look like. The authors have a section for each year from 1996 to 2055.

Williams, Robin C. *The Little iMac Book*. Berkeley, CA: Peachpit Press, 1998. 180 pages. ISBN: 0-201-35421-7.

The author of this book is a recognized expert on the Macintosh Computer system. The iMac represents the type of personal computer that Intel's former CEO Andy Grove has dubbed the future of computing: an easy-to-configure, simple-to-navigate, Internet appliance for the masses. The author takes time to explain the use of Apple's very successful machine while also providing a basic overview of the Web and digital communications.

Wilson, Mike. *The Difference Between God and Larry Ellison: Inside Oracle Corporation*. New York: William Morrow & Company, 1997. 385 pages. ISBN: 0688149251.

The difference between God and Larry Ellison, as the punchline explains, is "God doesn't think he's Larry Ellison." As the outspoken head of Oracle and the number one enemy of Microsoft and all things Bill Gates, Ellison's views and actions are important indicators on the digital frontier. The number two software maker is always worth watching because he may just try harder and become number one.

Wresch, William C. *Disconnected: Haves and Have-Nots in the Information Age.* Piscataway, NJ: Rutgers University Press, 1996. ISBN: 0-8135-2370-2.

This analysis of the social problems that may result as a consequence of how much access one has to information is written by a computer scientist. The author points out that it takes more than a modem and a computer to get the type of information that will do the most good given an individual's situation.

Yu, Albert. *Creating the Digital Future: The Secrets of Consistent Innovation at Intel.* New York: Simon & Schuster Trade, 1998. 265 pages. ISBN: 0-684-83988-1.

This book is an analysis of how the highly successful microprocessor company, Intel, has been able to take full advantage of founder Gordon Moore's "law." By combining technological innovation with a keen business acumen, Intel continues to double processing power per dollar value almost every 18 months.

JOURNALS

The following list includes both refereed and nonrefereed journals, industry newsletters, and a few important trade and consumer magazines. They are organized in alphabetical order, but the subject matter is an eclectic mix of technical writing on algorithms, microprocessor design, applications technologies, Internet and telecommunications, consumable products, and almost everything else digital.

ACM/IEEE Transactions on Networking (TON)
ACM Member Services
New York, NY
Web site: www.ton.cc.gatech.edu/

This is a co-sponsored journal of the ACM and the IEEE Computer and Communications Societies that offers coverage of research and reporting on network architecture and design, communication protocols, network software and technologies, services and applications, and network operations and management.

Artificial Life (Alife)
The MIT Press
Cambridge, MA
Web site: http://alife.santafe.edu/

The journal of the Santa Fe Institute is devoted to the investigation of scientific, engineering, philosophical, and social issues involving the synthesis of life-like behaviors in computers, machines, and other forms of artificial life.

Artificial Life and Robotics
International Symposium for Artificial Life and Robotics
Web site: link.springerny.com/link/service/journals/10015/index.htm

This print-only international journal was first published in 1998. It concentrates on publishing original technical papers and reviews on the development of new technologies involved in artificial life and robotics, especially computer simulations and digital hardware.

Communications of the ACM
Association for Computing Machinery (ACM)
New York, NY
Web site: www.acm.org/catalog/journals/101.html

Perhaps the premier international magazine covering all subjects in the field of computing, *Communications* is delivered monthly to members of the ACM. The journal covers industry news and emerging areas of computer technology, and is one of the most widely read and highly respected sources for accurate information in the field.

Computational Complexity (CC)
Birkhäuser Publishing, Ltd
Basel/Switzerland
Web site: link.springer.de/link/service/journals/00037/index.htm

In both print and electronic versions, this journal presents cutting-edge research in computational complexity, concentrating on the interface between mathematics and theoretical computer science. Main topics include models of computation, complexity bounds, and complexity classes.

Computer
Institute of Electrical and Electronics Engineers, Inc.
Piscataway, NJ
Web site: computer.org/computer/

Designed to provide more in-depth data than that found in the trade magazines, this is the flagship publication of the IEEE Computer Society. In their opinion, "*Computer* clearly defines the major trends in computer science and engineering without compromising technical quality. *Computer* has earned the loyalty and respect of computer scientists and engineers worldwide for more than 30 years. *Computer* is sent each month to all members of the IEEE Computer Society."

The Computer Journal
Oxford University Press
Web site: http://www.psyber.com/~tcj/

This official organ of the British Computer Society publishes research papers in a wide range of areas to provide a complete overview of developments in the field of computer science.

Computing
Springer-Wien, New York
New York, NY
Web site: link.springer.de/link/service/journals/00607/index.htm

Computing is an international journal for professionals and students in all fields of scientific computing. The journal presents the research results from computer science inquiry and computation experimentation. Discrete algorithms, symbolic computation, performance and complexity evaluation, operating systems, scheduling, software engineering, picture processing, parallel computation, classical numerical analysis, numerical software, numerical statistics, optimization, computer arithmetic, interval analysis, and plotting are some of the subjects presented in this print and electronic publication.

Computing and Visualization in Science
Springer-Verlag, Berlin/Heidelberg
Heidelberg, Germany
Web site: link.springer.de/link/service/journals/00791/index.htm

Available in electronic format to subscribers, this publication is devoted to computational sciences, a field of inquiry that brings together the multiple disciplines of mathematics, computer science, physics, chemistry, environmental sciences, biosciences, and engineering for the purpose of developing more sophisticated computer models.

Distributed Computing
Springer-Verlag, Berlin/Heidelberg
Heidelberg, Germany
Web site: link.springer.de/link/service/journals/00446/index.htm

This publication accepts reports on current implementations of distributed computing architectures and papers delineating theories on how future applications may develop. Concentration is on, but not limited to, the areas of novel architectures of distributed systems, architectures of computer networks, communication protocols, and distributed operating systems.

Dr. Dobb's Journal (DDJ)
Dr. Dobb's Journal
San Mateo, CA
Web site: www.ddj.com

Starting in 1976, *DDJ* has featured articles written by professional software developers covering every software language, platform, or tool. The journal is a well-respected source for new insights and timely reports on the happenings in software development. The companion Web site listed above includes extra features that enhance the printed version.

Forbes ASAP
Forbes Magazine, Inc.
New York, NY
Web site: www.forbes.com/asap

Forbes magazine publishes *ASAP* four times a year to specifically report on the newest computer technologies and how they will effect the commercial markets and society. It is a valuable resource because of its well-researched articles and list of respected columnists and analysts like George Gilder. Archives on the Web are located at the above Web site.

IEEE Annals of the History of Computing
Institute of Electrical and Electronics Engineers, Inc.
Piscataway, NJ
Web site: computer.org/annals/

Published under the IEEE Computer Society banner, this magazine seeks to preserve the rich history of the development of computer systems. Articles include interesting insights into the people, places, and circumstances that have blended to change the world as we know it.

IEEE Intelligent Systems
Institute of Electrical and Electronics Engineers, Inc.
Piscataway, NJ
Web site: computer.org/intelligent/

This quarterly journal of the IEEE Computer Society accepts articles of research and reporting in the area of intelligent systems development. For instance, the Nov/Dec 1998 issue is devoted to vision-based driving assistance. Members of the computer society can gain access to the articles online at the Web site listed above.

IEEE Internet Computing
Institute of Electrical and Electronics Engineers, Inc.
Piscataway, NJ
Web site: computer.org/internet/

Another publication of the IEEE Computer Society, this journal is one of the best for those hoping to keep up-to-date in the rapidly changing world defined by the ubiquitous Internet. The companion electronic site IEEE Internet Computing Online enhances the print version and is available at the Web site listed above to anyone with a browser.

IEEE Micro
Institute of Electrical and Electronics Engineers, Inc.
Piscataway, NJ
Web site: computer.org/micro/

Microcomputer and microprocessor systems are the focus of this IEEE Computer Society journal. The journal includes many topics related to the development of new digital systems and publishes helpful tutorials where applicable. Like many of the IEEE publications, full text access is available electronically for those members who are registered. The general public can often find free resources at the Web site listed above.

IEEE Software
Institute of Electrical and Electronics Engineers, Inc.
Piscataway, NJ
Web site: computer.org/software/

According to the IEEE Computer Society, which is responsible for this journal, *Software*'s "peer-reviewed articles, topical interviews, and columns by seasoned practitioners illuminate all aspects of the industry, including process improvement, project management, development tools, software maintenance, Web applications and opportunities, testing, and human factors."

The Information Society (TIS)
Taylor & Francis Publishers
London, England
Web site: www.taylorandfrancis.com

This peer-reviewed journal has been published since 1981 with the goal of publishing "leading edge analysis of the impacts, policies, system concepts, and methodologies related to information technologies and changes in society and culture."

International Journal of Computers and Their Applications
International Society for Computers and Their Applications
Cary, NC
Web site: www.isca-hq.org/journal.htm

The publishers describe their journal as a quarterly "with the purpose of providing a forum for state-of-the-art developments and research in the theory and design of computers as well as current innovative activities in the applications of computers. In contrast to other journals, this journal encourages contributions in real world applications of emerging computer technology, as well as advances in theoretical areas. Current topics of interest include, but are not limited to: architecture, networks, intelligent systems, parallel and distributed computing, software and information engineering, and computer applications (e.g., engineering, medicine, business, education, etc.)."

IT Professional
Institute of Electrical and Electronics Engineers, Inc.
Piscataway, NJ
Web site: computer.org/itpro/

This bimonthly publication of the IEEE Computer Society was first published in 1999 and is directed to an audience of developers and managers of information systems. New technologies, Internet businesses and applications, network security, data warehousing and mining, knowledge management and architectures, multimedia applications, artificial intelligence, operating systems,and software development are just a few of the areas of concern for this journal.

Journal of the ACM (JACM)
Association for Computing Machinery (ACM)
New York, NY
Web site: www.acm.org/jacm/

This refereed journal of the ACM presents theoretical research in the areas of complexity of algorithms, computer architecture, system modeling, artificial intelligence, data structures, database theory, and graph theory.

Looking.forward (A Supplement to *Computer*)
Institute of Electrical and Electronics Engineers, Inc.
Piscataway, NJ
Web site: computer.org/student/looking/

This publication is the IEEE Computer Society's student newsletter, publishing articles of interest for student members of that division of the IEEE. The archives of this newsletter are available to anyone online at the Web site listed above.

Machine Vision and Applications
Springer-Verlag, Berlin/Heidelberg
Heidelberg, Germany
Web site: link.springer.de/link/service/journals/00138/index.htm

This is a journal of machine vision research and development for which submissions are sought for reports on all applications and engineering aspects of image-related computing. Of particular interest are contributions on scientific, commercial, industrial, military, and biomedical applications of machine vision that report on algorithms, architectures, VLSI implementations, AI techniques and expert systems, front-end sensing, multidimensional and multisensor machine vision, real-time techniques, image databases, virtual reality and visualization.

Mobile Networks and Applications
Baltzer Science Publishers
Bussum, The Netherlands
Web site: www.cs.wisc.edu/~lists/archive/dbworld/0233.html

The sister journal of *Wireless Networks*, this publication is designed as a forum for the reporting on advances in mobility. Of special importance are the extra issues, which cover topics like nomadic computing; applications and services supporting the mobile user; design and analysis of algorithms for online and mobile environments; protocols to cope with mobility, limited bandwidth, intermittent connectivity, and mobile applications; and location-dependent and sensitive applications.

Soft Computing
Springer-Verlag, Berlin/Heidelberg
Heidelberg, Germany
Web site: link.springer.de/link/service/journals/00500/index.htm

In both print and electronic form, this journal is designed to publish important results on computing technologies, bringing together the research areas of evolutionary algorithms and genetic programming, neural science and neural net systems, fuzzy set theory and fuzzy systems, and chaos theory and chaotic systems.

Software—Concepts & Tools
Springer-Verlag, Berlin/Heidelberg
Heidelberg, Germany
Web site: link.springer.de/link/service/journals/00378/index.htm

This is an international journal of software systems. With a target audience of professional software developers, scientists, and computer science students, the journal concentrates on the theoretical, technical, and organizational foundations required for the production of software.

Theory of Computing Systems
Springer-Verlag, New York
New York, NY
Web site: link.springer.de/link/service/journals/00224/tocs/current.html

Formerly known as *Mathematical Systems Theory*, this online version of the respected international journal on computing theory can be found at the Web site listed above.

Virtual Reality
Springer-Verlag, London
London, England
Web site: link.springer.de/link/service/journals/10055/index.htm

This journal, as the name suggests, concentrates on the multidisciplinary topic of virtual reality (VR) design and implementation. Its articles deal with topics such as technology and software, human factors, applications, assessment of VR systems, and ethical issues.

The Visual Computer
Springer-Verlag, Berlin/Heidelberg
Heidelberg, Germany
Web site: www-ci.u-aizu.ac.jp/visualcomputer/

This title concentrates on the world of computer graphics. According to the publisher, this print and electronic journal's , " scope includes detection and communication of visual data, intermediate data structures and processing techniques for visual data and computer graphics, and graphical representations of images. In addition to the technical contributions, a regular column carries product news from both academia and industry. A calendar of events provides information on forthcoming meetings and symposia. "

Wired
Wired Digital, Inc.
San Francisco, CA
Website: www.wired.com/wired/ads/

Wired was the first magazine to see the importance of digital applications on modern society. While it is presented in a fast-paced, MTV style, the reporting is accurate, interesting, and an important indicator of the trends taking place in the new age.

Wireless Networks
Baltzer Science Publishers
Bussum, The Netherlands
Web site: www.baltzer.nl/winet/winet.asp

Reporting on a developing protocol that is changing the way data networking and telecommunications function, *Wireless Networks* focuses on architectures for personal communications, wireless LANs, radio, networks, etc.

WEB RESOURCES

The final section of this chapter is a compendium of Web-based resources that provide the latest information on the broad range of development in computer science. Should any of the links provided herein become broken, it is suggested that the readers use their favorite search engines to look up both the names and subject key words for the listing.

ADSL Forum
www.adsl.com

A telecommunications industry Web site that includes some useful information about current and future applications of Asymmetric Digital Subscriber Line (ADSL) services over the telephone lines.

Artificial Life Bibliography of On-line Publications
Ezequiel Di Paolo
www.cogs.susx.ac.uk/users/ezequiel/alife-page/alife.html

This online resource is kept up-to-date, making it one of the best sites for finding publications related to the field of artificial life. Nearly 500 publications are linked through this Web page. Among the publication links contained in this Web site are *Behavior and Social Behavior; Biology; Complexity; Self-Organization; Molecular Evolution; Development Evolution and Learning; Evolutionary Robotics; Implementations and Applied Artificial Life;* and *Philosophy, Methodology and Implications.*

Artificial Life Online
alife.santafe.edu

The Santa Fe Institute's online version of the print journal, *Artificial Life* "is devoted to a new discipline that investigates the scientific, engineering, philosophical, and social issues involved in our rapidly increasing technological ability to synthesize life-like behaviors from scratch in computers, machines, molecules, and other alternative media."

The Bulletin of the Santa Fe Institute
www.santafe.edu/sfi/publications/Bulletins/

Published by the Santa Fe Institute (SFI), mostly as a communiqué to its supporters, the bulletin is a free overview of the progress and plans undertaken by this prestigious center of computer modeling.

Career Resource Library
America's Career InfoNet
www.acinet.org/acinet/resource/occup/engin.htm#comp

This site as a whole is a clearinghouse for all types of careers in North America. The specific site referred to above has a link to computer-related occupations. A sample of some of the links (and the link sponsors) displayed on this page include *Computer Engineer,* New York State Department of Labor; *Computer Programmers,* British Columbia Occupational Outlooks; *Computer Specialists Occupations in Robotics and Automated Systems,* California Occupational Guide; *Computer Support Specialist,* New York State Department of Labor; *Computer Systems Analysts,* British Columbia Occupational Outlooks; *Database Administrator,* New York State Department of Labor; *Information Manager,* Princeton Review; *Multimedia Developers,* Computer Museum; *Software Developer,* Princeton Review; *System Analyst,* New York State Department of Labor; and *Web Page Designers, Masters, and Mistresses,* California Occupational Guide.

CMPNet
www.cmpnet.com

This is the main site for CMP Media, Inc., one of the premier publishers of computer industry magazines for both information technology professionals and the consuming public. From this one Web page, the reader will find links to all of the online versions of their publications. Among these are *BYTE* www.byte.com; *ChannelWeb* www.channelweb.com; *Data Communications* www.data.com; *EDTN Electronics Design Tech and News network* www.edtn.com; *EE Times electrical engineer news* www.eetimes.com; *InformationWeek* www.iweek.com; *InternetWeek* www.internetwk.com; *Network Computing* www.nwc.com; *Planet IT* www.planetit.com/home; *TechEncyclopedia* www.techweb.com/encyclopedia/?ls=ch_cw_te; *TechWeb* www.techweb.com/; *TechWeb Internet* www.techweb.com/internet; *tele.com* www.teledotcom.com; *UnixWorld* www.networkcomputing.com/unixworld; *WANsites* www.networkcomputing.com/wansites; *WebTools* www.webtools. com; *Win 98 Megasite* www.cmpnet.com/win98; *Windows Magazine* and www.winmag.com.

CNET
CNET, Inc., 1995-1999
http://www.cnet.com/

CNET is a highly respected first stop on the Web for links to the latest advances in commercial software, hardware, and computer product development. A unique element of CNET is a multimedia aspect that is responsible for video programming and radio shows—all accessible from the Web site.

Collection of Computer Science Bibliographies
Alf-Christian Achilles, 1994-1997
liinwww.ira.uka.de/bibliography

Containing nearly 1 million references from more than 1,200 bibliographies of scientific literature in computer science, this incredible online resource also provides countless hyperlinks to full text versions of the cited papers.

Complexity International
www.csu.edu.au/ci/

Complexity International is a refereed journal for research and reporting of complex systems. Writings in this journal deal with the interaction of multiple parallel processes to facilitate artificial life, cellular automata, chaos theory, control theory, evolutionary programming, fractals, genetic algorithms, neural networks, and other topics.

The Computer Bulletin
British Computer Society, Oxford University Press, 1998

www.bcs.org.uk/publicat/journals/bulletin/calendar/1998.htm

This online magazine for members of the British Computer Society has been in circulation for over 40 years. It serves as the main voice of the society to its members and the wider community. Freely accessible via the Web site listed above, it includes numerous articles covering many aspects of computer science and new digital applications. Archives of past issues are also available online.

Computer Information Centre (CompInfo)
The Computer Information Centre
www.compinfo.co.uk

An online source for all things computer related, this site is one of the best informational sites on the Web for current information. Some examples of category links are Anti-Virus Software; Apple Macintosh; Artificial Intelligence, Intelligent Agents; AS/400; Business Process Re-engineering; CAD; Chip and Semi-Conductor Technology; Computer Telephony Integration (CTI); Document Imaging and Management, Search Tools; EDI, Electronic Commerce, Smartcards; EIS, OLAP, DataWarehousing, Report Writers; Electronic Publishing, Desktop Publishing, Web Publishing; Groupware, E-mail, Workflow; Hardware generally; Intranets, Internet, and Network Computers (NC); Mobile Computing, PDAs; Multimedia; Networking and Data Communications; Robotics; Software Development; Supercomputers, Parallel Processors and High Performance Computing; System, Security, and Storage Management; UNIX, Aix, etc.; Video Conferencing, Desktop Conferencing; Virtual Reality, VRML; Voice/Speech Technology; Windows, Windows NT; and Year 2000 Date Problem.

Computer Magazine Archive
Ziff-Davis, Inc., ZDNet
cma.zdnet.com/texis/cma/cma

Claiming to have access to over 200,000 articles online, this Ziff-Davis sponsored site offers a searchable index of approximately 236 top computer and computer-related magazines, journals, newsletters, and newspapers. It is a highly recommended resource. A search by registered users will retrieve articles, or abstracts of articles, from many publications. As of February 1999, a sample of some of the publications available through this site are *Advertising Age, Byte, C/C++ Users Journal, CAD/CAM Update, Communications International, Computer Aided Design Report, Computer Business Review, Computer Industry Report, Computer Life, Computer Protocols, Computer Weekly, Consumer Electronics, Digital Media, Digital News & Review, dot.COM, Dr. Dobb's Journal, e-Business Advisor, Electronic Business Today, Emediaweekly, Fiber Optics News, HP Professional, Health Management Technology, Hewlett-Packard Jour-*

nal, Home PC, Information Outlook, Inside the Internet, Intelligent Enterprise, Inter@ctive Week, Internet Week, LAN Times, MacWEEK Online, Multimedia World, NDT Update, Newsbytes, PC Magazine, Tele-communications, The Futurist, The Seybold Report on Publishing Systems, U.S. News & World Report, UNIX News, Wall Street & Technology, Windows Magazine, and *Wireless World.*

Computerworld Research Links
Computerworld, Inc., IDG Enterprises
www.computerworld.com/res/research_links.html

Created by the publishers of one of the oldest computer industry magazines, *ComputerWorld,* this page offers the online version of their publication, which is loaded with many links to resources, papers, research results, and excellent up-to-date information on many computer science topics

CRA's Forsythe List
Computing Research Association
www.cra.org/Contacts/cscedb.html

This is a well-maintained searchable database containing contact information for and Web links to all Ph.D. granting computer science and computer engineering departments in the United States and internationally. It also contains listings of laboratories and research centers in the industry and the government. The list is named for Dr. George Forsythe, a professor of computer science at Stanford University and a founder of computer science graduate education.

CRA Taulbee Survey of Ph.D. Granting Institutions
Computing Research Association
www.cra.org/statistics/

Named for the man who conducted these yearly surveys from 1970–1984, Orrin E. Taulbee of the University of Pittsburgh, they are now in their 28th year. The survey is the main source of information on the enrollment, production, and employment of Ph.D.s in computer science and computer engineering for North America. It includes salary and demographic data with breakdowns according to gender and ethnicity. The data for the past academic year and several earlier years are available for free; download in PDF format at the Web site listed above.

Encryption FAQ
RSA Data Security, Inc.
www.rsa.com/rsalabs/faq/

This is the fourth version of RSA Laboratories' *Frequently Asked Questions about Today's Cryptography.* This FAQ covers the technical mathematics of cryptography as well as export law and basic fundamentals of information security. RSA is one of the world leaders in data security.

Free On-line Dictionary of Computing (FOLDOC)
Denis Howe
foldoc.doc.ic.ac.uk/

This resource is one of the most useful found on the Internet. Maintained by Denis Howe since 1985 "FOLDOC is a searchable dictionary of acronyms, jargon, programming languages, tools, architecture, operating systems, networking, theory, conventions, standards, mathematics, telecoms, electronics, institutions, companies, projects, products, history, in fact anything to do with computing." All terms listed in the glossary of this book originated on FOLDOC and are used with permission from Denis Howe.

Gilder Technology Report
George Gilder
www.gildertech.com/html/archives.html

Senior fellow of the Discovery Institute and president of Gilder Technology Group, Inc., George Gilder has a reputation as one of America's most intelligent technology/computer pundits. This site solicits a subscription to the current issue, but provides free access to archives of the past reports that cover many aspects of computer and communication development.

I, Cringely
Robert X. Cringely, PBS Online, and Oregon Public Broadcasting
www.pbs.org/cringely/index.html

Cringely is probably best known for his recent PBS series, "Triumph of the Nerds," but he was an expert analyst for the computer trade magazine, *InfoWorld* for eight years. He has been covering the industry for at least two decades. This online resource is available for free at the Web site listed above. On the site, Cringely often breaks news, makes predictions, and skewers the major players from Silicon Valley and Redmond. This is an excellent place to find out what is happening in commercial computer technology.

IDG.net
International Data Group (IDG)
www.idg.net

This is the corporate Web portal of the IDG, publisher of many computer industry magazines and books. From this site, the reader will have access to current articles chosen from *ComputerWorld*, *PC World*, *NetworkWorld*, and *Industry Standard*.

IEEE Internet Computing Guide to Web Resources
Institute of Electrical and Electronics Engineers, Inc.
computer.org/internet/links.htm

This site of the IEEE magazine Internet Computing lists scores of Web accessible resources that can help the researcher learn more about developments in computer science and provides free links. Infrastructure, hardware, software, networking, and associations are just a few of the categories available to anyone with a browser.

IEEE Internet Computing Online
Institute of Electrical and Electronics Engineers, Inc.
computer.org/internet/

Published under the IEEE Computer Society banner, this online version of the IEEE Internet Computing journal is one of the best to read for those hoping to keep up-to-date in the rapidly changing world defined by the ubiquitous Internet. *IEEE Internet Computing Online* enhances the print version and is freely available to anyone with a browser. As the Computer Society page says, this site publishes "the latest developments in Internet-based applications and supporting technologies and addresses the Internet's widening impact on engineering practice and society."

The Information Society
www.slis.indiana.edu/TIS/tisbib.html

This is a comprehensive bibliography of monographs and papers related to the ways information technology affects society. Some abstracts are available online. The site is maintained by the international IT journal *The Information Society*, "published since 1981, [it] is a key critical forum for leading edge analysis of the impacts, policies, system concepts, and methodologies related to information technologies and changes in society and culture."

Intelligent Data Analysis
Elsevier Science, Inc.
www-east.elsevier.com/ida/

Intelligent Data Analysis is a paper and Web-based journal that publishes papers related to the research and applications of artificial intelligence techniques. According to the publisher, "these techniques include (but are not limited to): all areas of data visualization, data pre-processing (fusion, editing, transformation, filtering, sampling), data engineering, database mining techniques, tools and applications, use of domain knowledge in data analysis, machine learning, neural nets, fuzzy logic, statistical pattern recognition, knowledge filtering, and post-processing. In particular, we prefer papers that discuss development of new AI architectures, methodologies, and techniques and their applications to the field of data analysis."

Journal of Artificial Societies and Social Simulation (JASSS)
SIMSOC Consortium
www.soc.surrey.ac.uk/JASSS/1/4/contents.html

This is "an interdisciplinary journal for the exploration and understanding of social processes by means of computer simulation."

LIDO Telecom WebCentral
Lillian Goleniewski, LIDO Organization
www.telecomwebcentral.com/

One of the best online resources for information related to the subject of telecommunications and digital connectivity. Categories (and number of links) at the end of 1999 included the following: Books (357), Career Search (40), Career Conferences (63), Distributed Virtual Environments (25), Economics and Finance (176), Encyclopedias and Virtual Libraries (27), Glossaries (35), Industry Stock Quotes (522), Industry and User Associations (435), Intelligent Agents (18), Legal and Regulatory (137), Magazines (124), Market Research Sources (22), Newsgroups (2), Power Utilities and Telecom (24), Standards Bodies (53), Technology Sites (1,002), TeleEducation (79), Telecom Internet and IT Companies (1,968), Telecom Museums (19), Telecom News (38), Telecom Studies (24), Telemedicine (58), and Web Development (120).

Losing Ground Bit by Bit
Susan Goslee with Chris Conte, Benton Foundation
www.benton.org/Library/Low-Income/

Produced in association with the National Urban League, this report looks at the challenges faced by low-income Americans as their country embraces the technologies of the information age. This is an excellent review of the issues related to technology "haves and have-nots."

National Newsbytes
Association of Women in Computing (AWC)
http://www.awc-hq.org/publications/index.html

This is the online archive of the AWC national newsletter containing current organization news and news briefs about women and computers.

P a r a S c o p e
Institute of Electrical and Electronics Engineers, Inc.
computer.org/parascope/

Maintained by the IEEE Computer Society, this Web page provides links to parallel computing sites. A partial list of links includes supercomputing centers and parallel computing, HPC journals, HPC conferences, high-performance computing vendors, federal agencies, TOP500 supercomputer sites, and a list of the world's most powerful computing sites.

PCWorld Online
PC World Communications
www.pcworld.com

The print magazine *PCWorld* is one of the more popular resources for desktop computer news. This online resource, *PCWorld Online* enhances that publication with many freely available resources, articles, and links for the interested reader.

Phil Agre's Home Page
dlis.gseis.ucla.edu/people/pagre/

A very interesting starting point to resources about communication technologies with an emphasis on how new digital methods are changing our institutions and ourselves. Maintained by Phil Agre, associate professor of information studies at the University of California, Los Angeles, the page includes links to many of his recent writings including *The Architecture of Identity: Embedding Privacy in Market Institutions*, Proceedings of the Telecommunications Policy Research Conference, Alexandria, Virginia, October 1998; *The Internet and Public Discourse, First Monday* 3(3), 1998; *The Next Internet Hero, Technology Review*, November 1997, page 61; *Building an Internet Culture, Revista USP* (University of Sao Paulo) 35, 1997, pages 112-117; *Computation and Human Experience*, Cambridge University Press, 1997 (edited with Marc Rotenberg); and *Technology and Privacy: The New Landscape*, MIT Press, 1997.

Agre has also included a regularly updated bibliography of books on the social aspects of computing, which is available at dlis.gseis.ucla.edu/people/pagre/recent-books.html.

The Problem You May Not Know You Have
Paul Gillin, *ComputerWorld*
www.flash.net/~bschoen/article.html

This is the reprint of the first article to mention the Y2K bug. The author, Paul Gillin profiled systems analyst William Schoen who is credited as the first computer professional to raise the alarm about the chaos that the programming glitch would cause.

Timeline of Computer History
IEEE Computer Society
computer.org/computer/timeline/

This is a freely available timeline of significant events in the development of computers and digital systems. The Computer Society makes this file available for download in Adobe PDF format.

UniGuide Academic Guide to the Internet
Aldea Communications, Inc.
www.aldea.com/guides/ag/ciscats.html

The guide is created and maintained especially for the higher education community with the goal of developing the primary Internet resource of and for the research and academic community. An interesting element includes a feedback component that allows for the independent rating of included resources. For the purposes of research in computer science, pertinent sections are advanced scientific computing; computer and computation research; information, robotics and intelligent systems; microelectronic information processing systems; and networking and communications research and infrastructure.

UNIX Reference Desk
www.geek-girl.com/unix.html

This document contains references to material from a number of different sources on computing in the UNIX environment. The resources in this document are divided into general, Texinfo pages, applications, programming, IBM AIX systems, HP-UX systems, Unix for PCs, Sun systems, X Window system, networking, security, and humor classes.

whatis.com
whatis.com, Inc.
www.whatis.com

This is a searchable electronic encyclopedia that covers many aspects of computer science and the computer business. Freely available, the publishers also include learning path features like "How the Internet Works," to aid research and facilitate search options for its users.

xDSL.com
TeleChoice, Inc.
www.xdsl.com

This site is dedicated to providing pertinent current information about the deployment of digital subscriber line services in all of its many flavors. A breaking xDSL news section includes current events and announcements. While the reader should be aware that TeleChoice is making conscious decisions about what to provide its readers, there is some very valuable information at this Web site. Of particular value is a search engine that will recover articles on any computer-related topic from a host of reputable content providers.

Y2K: So Many Bugs ... So Little Time
Scientific American
www.sciam.com/1999/0199issue/0199dejager.html

A clear and simple analysis of the Y2K problem and its resolution. Written by the man who sounded the alarm in a *ComputerWorld* article in 1993.

ZDNet Anchordesk
Ziff-Davis, Inc., ZDNet
www.zdnet.com/anchordesk/

Edited by the highly regarded computer industry analyst, Jesse Berst, this daily Web site is a constantly-upgraded well of information about emerging trends, new products, potential problems, and advice about a segment of the economy and society that is changing faster than anyone can comprehend.

CHAPTER ELEVEN
Glossary

The following is a glossary of frequently used computer terms that may have been used in this book. The contributions in this chapter are excerpted and/or paraphrased from an exceptional resource called *The Free On-line Dictionary of Computing* (FOLDOC). FOLDOC is a searchable archive of over 12,000 computer-related terms compiled from contributions of over 900 people worldwide. FOLDOC has been edited, maintained, and copyrighted from 1993 to 1999 by Denis Howe, who began the project in 1985. The latest version of FOLDOC is accessible at http://foldoc.doc.ic.ac.uk/. (The master site is http://wombat.doc.ic.ac.uk/.) The portion reprinted here is used with permission from Denis Howe.

Out of respect for the author(s), an attempt has been made to reproduce this work verbatim, but some definitions have been edited for length, spelling (mostly British to U.S. nomenclature), or factual representation and accuracy.

ADSL (Asymmetric Digital Subscriber Line or Asymmetric Digital Subscriber Loop)   A form of digital subscriber line in which the bandwidth available for downstream connection is significantly larger then for upstream. Although designed to minimize the effect of crosstalk (interference between competing signals) between the upstream and downstream channels, this setup is well suited for Web browsing and client-server applications as well as for some emerging applications such as video on demand.

The data-rate of ADSL strongly depends on the length and quality of the line connecting the end-user to the telephone company. Typically the upstream data flow is between 16 and 640 kilobits per second (kbps) while the downstream data flow is between 1.5 and 9 megabits per second (Mbps). ADSL also provides a voice channel. ADSL can be configured to carry digital data, analog voice, and broadcast MPEG2 video in a variety of implementations to meet customer needs.

algorithm   A detailed sequence of actions that perform to accomplish some task. Named after an Iranian mathematician, Al-Khawarizmi. Technically, an algorithm must reach a result after a finite number of steps, thus ruling out brute force search methods for certain problems, though some might claim that brute force search was also a valid (generic) algorithm. The term is also used loosely for any sequence of actions that may or may not terminate.

architecture   Design, or the way components fit together. The term is used particularly for processors, both individual and in general. ("The ARM has a really clean architecture.") It may also be used of any complex system, e.g., "software architecture" or "network architecture."

artificial intelligence (AI)   The subfield of computer science concerned with the concepts and methods of symbolic inference by computer and symbolic knowledge representation for use in making inferences. AI can be seen as an attempt to model aspects of human thought on computers. It is also sometimes defined as trying to solve by computer any problem that a human can solve faster.

bandwidth   1. The difference between the highest and lowest frequencies of a transmission channel. 2. The amount of data that can be sent through a given communications circuit per second.

baud   The unit in which the information carrying capacity or "signaling rate" of a communication channel is measured. One baud is one symbol (state-transition or level-transition) per second. This coincides with bits per second (bps) only for two-level modulation with no framing or stop bits. A symbol is a unique state of the communication channel, distinguishable by the receiver from all other possible states. For example, it may be one of two voltage levels on a wire for a direct digital connection or it might be the phase or frequency of a carrier.

The term baud was originally a unit of telegraph signaling speed, set at one Morse code dot per second. Generally, the term baud is the reciprocal of the duration of the shortest signaling element. Baud was proposed at the International Telegraph Conference of 1927, and named after J.M.E. Baudot (1845–1903), the French engineer who constructed the first successful teleprinter.

The term baud causes much confusion and is usually best avoided. Use bits per second (bps), bytes per second (Bps), or characters per second (cps) instead.

bus   A set of conductors (wires, PCB tracks, or connections in an integrated circuit) connecting the various functional units in a computer. There are busses both within the CPU and connecting the CPU to external memory and peripheral devices. The width of the bus (i.e. the number of parallel connectors) determines the size in bits of the largest data item it can carry. The bus width and the number of data items per second it can transmit are one of the factors limiting a computer's performance. Most current microprocessors have 32-bit busses both internally and externally.

cable modem   A type of modem that allows people to access the Internet via their cable television service. However, industry pundits often point out that the cable system still does not have the bandwidth or service level in many areas to make this feasible. For example, it has to be capable of two-way communication. A cable modem can transfer data at 500 kilobits per second (kbps) or higher, compared with 56 kbps for common telephone line modems, but the actual transfer rates in practice may be lower depending on the number of other simultaneous users on the same cable.

CD (Compact Disc)   (Not "disk," this spelling is part of the industry standard.) A 4.72-inch disc developed by Sony and Philips that can store, on the same disc, still and/or moving images in monochrome and/or color; stereo or two separate sound tracks integrated with and/or separate from the images; and digital program and information files. The same fabrication process is used to make both audio CDs and CD-ROMs for storing computer data. The only difference between CDs and CD-ROMs is in the device used to read the disk (the player or drive).

CD-R (Compact Disc Recordable)   A write-once version of CD-ROM. CD-Rs can hold about 650 megabytes (MB) of data. They are very durable and can be read by normal CD-ROM drives, but once data has been written it cannot be altered.

CD-ROM (Compact Disc Read-Only Memory)   A nonvolatile optical data storage medium using the same physical format as audio compact discs, readable by a computer with a CD-ROM drive. CD-ROM is popular for distribution of large databases, software, and especially multimedia applications. The maximum capacity of a CD-ROM is about 600 megabytes (MB). A CD can store around 640 MB of data—about 12 billion bytes per pound weight. CD-ROM drives are rated with a speed factor relative to music CDs (1x or 1-speed which gives a data transfer rate of 150 kilobytes per second (KBps)). 24x drives are common in 2000.

CMOS (Complementary Metal Oxide Semiconductor)   A semiconductor fabrication technology using a combination of n- and p-doped (negative- and positive-doped) semiconductor material to achieve low power dissipation.

CPU (central processing unit)   The part of a computer that controls all the other parts. Designs vary widely but, in general, the CPU consists of the control unit, the arithmetic and logic unit (ALU), and memory (registers, cache, RAM, and ROM), as well as various temporary buffers and other logic. The control unit retrieves instructions from memory and decodes them to produce signals. These signals control the other parts of the computer and may cause it to transfer data between memory and ALU, or to activate peripherals to perform input or output. A parallel computer has several CPUs that share other resources such as memory and peripherals.

data bus   The connections between and within the CPU, memory, and peripherals used to carry data. Other connections are the address bus and control
signals. The width of the data bus is one of the main factors determining the
processing power of a computer. Most current processor designs use a 32-bit
bus, meaning that 32 bits of data can be transferred at once. Some processors
have an internal data bus which is wider than their external bus in order to make
external connections cheaper while retaining some of the benefits in the
processing power of a wider bus.

data processing   The input, verification, organization, storage, retrieval,
transformation, and extraction of information from data. The term is normally
associated with commercial applications such as stock control or payroll.

distributed memory   The kind of memory in a parallel processor where each
processor has fast access to its own local memory and where in order to access
another processor's memory it must send a message via the inter-processor
network. The opposite of distributed memory is shared memory.

DVD (Digital Versatile Disc, formerly Digital Video Disc)   An optical storage
medium with improved capacity and bandwidth over the compact disc. DVDs,
like CDs, were initially marketed for home video entertainment. They became
available for computer users in late 1996. Today, DVD is making inroads as the
popular medium for full-length movies.

dynamic random access memory (DRAM)   A type of semiconductor memory
in which the information is stored in capacitors on an integrated circuit. Typically
each bit is stored as an amount of electrical charge in a storage cell consisting
of a capacitor and a transistor. Due to leakage the capacitor discharges
gradually and the memory cell loses the information. Therefore, to preserve the
information, the memory has to be refreshed periodically. Despite this inconvenience, the DRAM is a very popular memory technology because of its high
density (packing more bits per space) and consequent low price.

fiber optics   A plastic or glass (silicon dioxide) fiber no thicker than a human
hair used to transmit information using infrared or even visible light (usually a
laser) as the carrier. Optical fiber is less susceptible to external noise than other
transmission media, and is cheaper to make than copper wire, but it is much
more difficult to connect. Fiber optics are broadband carriers of information
between nodes on a network.

Fibre Channel-Arbitrated Loop (FC-AL or Fibre Channel)   A high-speed
serial bus interface standard intended to replace small computer system
interface (SCSI) on high-end servers. FC-AL has a number of advantages over
SCSI. First, it offers higher speeds than SCSI. The base speed of FC-AL is 100
mega bits per second (Mbps), which is likely to increase to 200, 400, and 800
Mbps in the next few years. Secondly, many devices are dual ported, meaning
they can be accessed through two independent ports, which doubles speed
and increases fault tolerance. Thirdly, cables for FC-AL can be as long as 30
meters (coaxial) or 10 kilometers (optical). Fourthly, FC-AL enables self-
configuring, hot swapping (connecting and disconnecting with power on), and a

maximum of 126 devices on a single port. Finally, FC-AL provides software compatibility with SCSI.

Despite all these features FC-AL is unlikely to appear on desktops anytime soon, partly because of its price and partly because typical desktop computers would not take advantage of many of its advanced features. On desktop systems, FireWire (defined under IEEE 1394) has more potential.

FireWire (High Performance Serial Bus)   See IEEE 1394.

General Public License (GPL)   The license applied to most software from the Free Software Foundation and the GNU project and other authors who choose to use it. The licenses for most software are designed to prevent users from sharing or changing it. By contrast, the GNU General Public License is intended to guarantee the freedom for all users to share and change free software. The GPL is designed to ensure that all license holders can distribute copies of free software (as well as charge for their distribution service if they wish); receive the source code for the software; and change the software or use pieces of it in new free programs. The GPL also strives to make license holders aware of the advantages of their licensing practice. The GPL forbids anyone to deny others these rights or to ask them to surrender these rights. These restrictions translate to certain responsibilities for those who distribute copies of the software or modify it.

GFLOPS (gigaflops, giga floating point operations per second)   One thousand million (10^9) floating point operations per second. One of them is strictly "one gigaflops"; it retains the s because it stands for *second* and not for the plural form. Floating point operations are those highly involved mathematical statements that are represented by notations that include a mantissa and their exponential values, as in the above notation of ten to the ninth power (10^9). Usually, microprocessors include an integrated circuit that handles these operations separately, speeding the total time it takes to solve exponential equations.

gigaPoP (giga Point of Presence) See PoP.

GNU /g*noo/   The Free Software Foundation's project to provide a freely distributed replacement for Unix. The GNU Manifesto was published in the March 1985 issue of *Dr. Dobb's Journal* but the GNU project started a year and a half earlier when Richard Stallman was trying to get funding to work on his freely distributable editor, Emacs. Emacs and the GNU C Compiler (GCC), are two tools designed for this project that have become very popular. GNU software is available from many GNU archive sites.

IEEE 1394 (FireWire)   A 1995 Macintosh/IBM PC serial bus interface standard offering high-speed communications and isochronous real-time data services. IEEE 1394 can transfer data between a computer and its peripherals at 100, 200, or 400 megabits per second (Mbps), with a planned increase to 2 gigabits per second (Gbps). Cable length is limited to 4.5 meters but up to 16 cables can be daisy-chained yielding a total length of 72 meters.

IEEE 1394 can daisy-chain together up to 63 peripherals in a tree-like structure (as opposed to SCSI's linear structure). It allows peer-to-peer device communication, such as communication between a scanner and a printer, to take place without using system memory or the CPU. It is designed to support plug-and-play and hot swapping. Its 6-wire cable is not only more convenient than the SCSI cables but can supply up to 60 watts of power, allowing low-consumption devices to operate without a separate power cord.

input device   A peripheral used to transfer data from the outside world into a computer system. Some input devices are operated directly by the user (via keyboard, mouse, touch screen, joystick, digitizing tablet, or microphone), while others are sensors or transducers that convert external signals into data such as using an analog to digital converter (this would also be true of a microphone). Other kinds of inputs are really one-half of a bidirectional link with another computer or storage device, such as a serial line or SCSI interface.

integrated circuit (IC or chip)   A microelectronic semiconductor device consisting of many interconnected transistors and other components.

Java   A simple, object-oriented, distributed, interpreted, robust, secure, architecture-neutral, portable, general-purpose programming language developed by Sun Microsystems. Java supports programming for the Internet in the form of platform-independent Java applets. Java programs can also run stand-alone on small computers. It is a write once, run anywhere programming language.

machine learning   The ability of a machine to improve its performance based on previous results. Neural networks are one kind of machine learning.

main memory   The storage device used by a computer to hold the currently executing program and its working data. A modern computer's main memory is built from RAM (random access memory) integrated circuits. Originally, ferrite core memory was a popular form of main memory, leading to the use of the term "core" for main memory. Computers have several other sorts of memory, distinguished by their access time, storage capacity, cost, and the typical lifetime or rate of change of the data they hold. Examples of memory/memory media are registers, cache, PROM (programmable read-only memory), magnetic disk (which may be used for virtual memory), and magnetic tape.

microprocessor   A computer whose entire CPU is contained on one (or a small number of) integrated circuits. The important characteristics of a microprocessor are the widths of its internal and external address and data busses (and their instructions), its clock rate, and its instruction set. The first commercial microprocessor was the Intel 4004 which appeared in 1971. Today's microprocessors run on 32-bit and 64-bit architecture. The PowerPC and Pentium are two common types.

natural language   A language spoken or written by humans, as opposed to a language used to program or communicate with computers. Understanding natural language is one of the greatest obstacles in using artificial intelligence because of the complexity, irregularity, diversity, and philosophical problems of interpreting the meaning of human language.

neural network   A network of very simple processors (units or neurons), each possibly having a small amount of local memory. The units are connected by unidirectional communication channels or connections, which carry numeric (as opposed to symbolic) data. The units operate only on their local data and on the inputs they receive via the connections.

A neural network consists of processing devices, either algorithms or actual hardware, whose design is inspired by the design and function of an animal brain and components thereof. Most neural networks have some sort of training rule whereby the weights of connections are adjusted on the basis of presented patterns. In other words, neural networks "learn" from example and exhibit some structural capability for generalization, just like children learn to recognize all dogs by seeing examples of a few dogs.

object-oriented programming (OOP)   The process of writing programs using programming languages and techniques based on the concept of an object or data structure (abstract data type). These objects are operated on using a set of routines, called methods. Operations on the data can only be performed using the same set of methods, which are common to all objects of a particular class. Since the same method is used for all operations, the interface between the operation and the objects is well defined, allowing the code implementing the methods to be changed so long as the interface remains the same. A major benefit of OOP is that code doesn't have to be rewritten over and over.

operating system (OS)   The low-level software that schedules tasks, allocates storage, handles the interface to peripheral hardware, and presents a default interface to the user when no application program is running. The OS may be split into a "kernel which is always present in the software. Various system programs use facilities provided by the kernel to perform higher-level house-keeping tasks, often acting as servers in a client-server relationship. Many programs include a graphical user interface and a window system as part of the OS.

The facilities and general design philosophy provided by an OS are extremely influential on the programming style and the technical cultures that emerge around the machines it runs. Example operating systems include 386BSD, AIX, AOS, Amoeba, Angel, Artemis microkernel, Brazil, COS, CP/M, CTSS, Chorus, DACNOS, DOSEXEC 2, GCOS, GEORGE 3, GEOS, ITS, KAOS, LynxOS, MPV, MS-DOS, MVS, Mach, Macintosh operating system (MAC OS), MINIX, Multics, Multipop-68, Novell NetWare, OS-9, OS/2, Pick, Plan 9, QNX, RISC OS, STING, System V, System/360, TOPS-10, TOPS-20, TRUSIX, TWENEX, TYMCOM-X, Thoth, Unix, VM/CMS, VMS, VRTX, VSTa, VxWorks, WAITS, Windows 95, Windows 98, and Windows NT.

PoP (Point of Presence)   A site where there exists a collection of telecommunications equipment, usually modems, digital leased lines, and multi-protocol routers. An Internet access provider may operate several PoPs distributed throughout a given area of operation to increase the chance that their subscribers will be able to connect to one with a local telephone call.

programming language   The formal language in which computer programs are written. The definition of a particular language consists of both syntax (how the various symbols of the language may be combined) and semantics (the meaning of the language constructs). Common programming languages include FORTRAN, Basic, Visual Basic, and Java.

protocol   A set of formal rules describing how to transmit data, especially across a network. Low-level protocols define the electrical and physical standards to be observed, such as bit- and byte-ordering, transmission of the bit stream, and error detection and correction of the bit stream. High-level protocols deal with the data formatting, such as the syntax of messages, the terminal to computer dialogue, character sets, sequencing of messages, etc.

scalability   How well a solution to a problem will work when the size of the problem increases.

semiconductor   A material, typically crystalline, that allows current to flow under certain circumstances. Common semiconductors are silicon, germanium, and gallium arsenide. Semiconductors are used to make diodes, transistors, and other basic solid state electronic components. As crystals of these materials are grown, they are doped or enhanced with traces of other elements called donors or acceptors to make regions that are n- or p-type, respectively for each electron model, or p- or n-type under the whole model. Where n- and p-type regions adjoin, a junction is formed that passes current in one direction (from p to n) but not the other, creating a diode.

shared memory   1. Memory in a parallel computer, usually RAM, which can be accessed by more then one processor, usually via a shared bus or network. It usually takes longer for a processor to access shared memory than to access its own private memory because of contention for the processor-to-memory connections and because of other overheads associated with ensuring synchronized access. Computers using shared memory usually have some kind of local cache on each processor to reduce the number of accesses to shared memory. This requires a cache consistency protocol to ensure that one processor's cached copy of a shared memory location is invalidated when another processor writes to that location.

The alternative to shared memory is message passing where all memory is private to some particular processor and processors communicate by sending messages down special links. This is usually slower than shared memory but it avoids the problems of contention for memory and can be implemented more cheaply.

2. Memory that can be accessed by more than one process in a multitasking operating system with memory protection. Some Unix variants, e.g., SunOS provide this kind of shared memory.

silicon   The material used as the base (or substrate) for most integrated circuits.

16 bit   This term often refers to the number of bits (binary storage units that can hold one of two values, such as 0 and 1 or true and false) used internally by a computer's CPU (central processing unit). ("The Intel 8086 is a 16-bit processor." Its external data bus or address bus may be narrower.) The term may also refer to the size of an instruction in the computer's instruction set or to any other item of data.

64 bit   Describes computer architecture based around the ALU (arithmetic and logic unit), registers, and data bus that are each 64 bits wide. 64-bit processors have been quite common today, e.g. Digital Alpha, versions of Sun SPARC, MIPS, and IBM AS/4000. The PowerPC and Intel are expected to move to 64 bits at their next generation—code named PPC 620 and Intel Merced.

A 64-bit address bus allows the processor to address 18 million gigabytes (GB) as opposed to the mere 4 GB allowed with 32 bits.

speech recognition (voice recognition)   The identification of spoken words by a machine. The spoken words are digitized (turned into a sequence of numbers) and matched against coded dictionaries in order to identify the words.

Most systems must be trained or programmed, requiring samples of all the actual words to be spoken by the user of the system. The sample words are digitized, stored in the computer, and used to match against future words. More sophisticated systems require a few voice samples, but not of every word. The more sophisticated systems use the voice samples in conjunction with dictionaries of larger vocabularies to match the incoming words. Future systems aim to be speaker-independent, meaning they will recognize words in their vocabulary from any speaker without training.

32-bit application   Designed as a result of the incompatibility of MS-DOS and Microsoft Windows, and Intel 8088 and 80286 microprocessors. These are 16-bit microprocessors with a segmented address space. Programs such as MS-DOS and Microsoft Windows that had more than 64 kilobytes (KB) of code and/or data therefore had to switch between segments quite frequently. As this operation is quite time consuming in comparison to other machine operations, the application's performance suffered. Furthermore, programming with segments is more involved than programming in a flat address space, giving rise to some complications in programming languages like memory models in C and C++.

The shift from 16-bit software to 32-bit software on IBM PC clones became possible with the introduction of the Intel 80386 microprocessor. This microprocessor and its successors support a segmented address space with 16-bit and 32-bit segments (or more precisely, segments with 16- or 32-bit address offset) or a linear 32-bit address space. For compatibility reasons, however, much of the software is still written in 16-bit models.

transistor   A three terminal semiconductor amplifying device that is the fundamental component of most active electronic circuits, including digital electronics. The transistor was invented on 23 December 1947 at Bell Laboratories. There are two kinds of transistors, the bipolar transistor (also called the junction transistor), and the field effect transistor (FET).

Unix   An interactive, time-sharing operating system (OS) invented in 1969 by Ken Thompson after Bell Laboratories left the Multics project. Thompson originally invented Unix so he could play games on his scavenged PDP-7 computer. Dennis Ritchie, the inventor of C, is considered a co-author of Unix. The turning point in Unix's history came when it was reimplemented almost entirely in C from 1972 to 1974, making it the first source-portable OS. Unix subsequently underwent mutations and expansions at the hands of many

different people, resulting in a uniquely flexible and developer-friendly environment.

By 1991, Unix became the most widely used multiuser general-purpose OS
in the world. Many people consider the popularity of Unix to be the most
important victory yet of hackers over industry opposition.

USB (Universal Serial Bus)   A standard promoted by Intel for communication
between a PC and external peripherals over an inexpensive cable using biserial
transmission. USB works at 12 megabits per second (Mbps) with specific cost
consideration for low cost peripherals. It supports up to 127 devices and both
isochronous (occuring at regular intervals) and asynchronous (occuring at
unregulated time intervals) data transfers. Cables can be up to 5 meters long
and include built-in power distribution for low power devices. The USB supports
daisy chaining (hooking together end to end) through a tiered star multidrop
topology.

Before March 1996 Intel started to integrate the necessary logic into PC
chip sets and encouraged other manufacturers to do the same. Widespread
availability and support is now common in new PCs. Because of its relatively low
speed, USB is intended to replace existing serial ports, parallel ports, keyboard
and monitor connectors, and keyboard compatible mice, monitors, printers, and
possibly some low-speed scanners and removable hard drives.

World Wide Web (WWW, W3, the Web)   An Internet client-server hypertext
distributed information retrieval system which originated from the CERN High-
Energy Physics laboratories in Geneva, Switzerland. On the Web everything
(documents, menus, indices) is represented to the user as a hypertext object in
HTML (hypertext markup language) format. Hypertext links refer to other
documents by their URLs. These can refer to local or remote resources accessible via FTP, Gopher, Telnet, or news, as well as those available via the http
(hypertext transfer protocol) used to transfer hypertext documents.

The client program or browser runs on the user's computer and provides
two basic navigation operations, following a link or sending a query to a server.
A variety of client and server software is freely available.

xDSL (Digital Subscriber Line, DSL, or Digital Subscriber Loop)   A family of
digital telecommunications protocols designed to allow high-speed data
communication over the existing copper telephone lines between end-users and
telephone companies (telco).

When two conventional modems are connected through the telephone
system (PSTN), xDSL treats the communication the same as a voice conversation. The advantage to this type of communication is that there is no investment
required from the telephone company. The disadvantage is that the bandwidth
available for the communication is the same as that available for voice conversations, usually 64 kb/s (DS0) at most. This is a disadvantage because twisted-pair
copper cables in individual homes or offices can usually carry significantly more
than 64 kb/s but the telco needs to handle the signal as digital rather than
analog.

INDEX

ABOUT THE AUTHOR

Martin K. Gay is a writer, computer consultant, and teacher concentrating in education and communication technologies. He has also worked as a consultant for various educational initiatives, including the Enhanced Learning through Electronic Communities (ELTEC) Project, Watershed Action through Education and Research (WATER) Neighbors, the Washington State University Extension Learning Center, the Jefferson County Higher Education Access Committee, the Evergreen Volunteer Tech Team, COM3, and Group Exploring the National Information Infrastructure (GENII). Gay is the author of *The New Information Revolution: A Reference Handbook.*